The Cambridge Nature Study Series

General Editor : Hugh Richardson, M.A.

THE GATEWAYS OF KNOWLEDGE

THE GATEWAYS OF KNOWLEDGE

AN INTRODUCTION TO THE STUDY OF THE SENSES

BY

J. A. DELL, M.Sc. (Vict.)

ASSISTANT MASTER, SEXEY'S SCHOOL, BRUTON, SOMERSET

Cambridge :

at the University Press

1912

CAMBRIDGE
UNIVERSITY PRESS

University Printing House, Cambridge CB2 8BS, United Kingdom

Published in the United States of America by Cambridge University Press, New York

Cambridge University Press is part of the University of Cambridge.

It furthers the University's mission by disseminating knowledge in the pursuit of education, learning and research at the highest international levels of excellence.

www.cambridge.org
Information on this title: www.cambridge.org/9781107655836

© Cambridge University Press 1912

First published 1912
First paperback edition 2013

A catalogue record for this publication is available from the British Library

ISBN 978-1-107-65583-6 Paperback

PREFACE

MR DELL has written on the Study of the Senses, on simple experimental psychology. Nature Study has been praised as a training in observation, and as, in the use of the telescope, the astronomer must study the error of the instrument, so, where that instrument is the naked eye, the naturalist must learn what tricks it may play him. Can we call anyone a trained observer who is unconscious of such errors as might arise from persistence of vision, from the negative after-image or from the blind spot of the retina?

That part of Nature which is most accessible for exploration in a school classroom is human nature—boy nature—"The proper study of mankind is man." Here, Mr Dell has shown us a proper way—the scientific way—of studying the mind of boy.

Hitherto, the laboratory psychologist has often regarded the schoolmaster as too untrained and too ignorant to be a competent ally as an experimenter in mental fields. Nor has the schoolmaster been altogether willing to regard his classroom as chiefly an exploration ground for the rambling psychological explorer. But now these studies are beginning to interest the rising generation of schoolmasters. If some of us were not so busy organizing laboratories and propagating cookery recipes for oxygen and chlorine, we might have leisure to explore the material lavished around us in the minds of our pupils.

Mr Dell has been fortunate in throwing his experiments into a form in which they can be used for classroom work. This often involves a great simplification and cheapening of apparatus. The methods have to be transformed from the conditions of the physician's consulting room where patients are elaborately treated one at a time to those of the classroom where boys are dealt with by the dozen. It is just this preliminary examination by the schoolmaster which may correctly select those who should be passed on for further examination by the school doctor.

Mr Dell has also shown how boys may be not merely passive under examination but active co-operators with their teacher in explorations of mutual interest. Many teachers have found the difficulty of organizing practical work for large classes. This book, therefore, while nominally showing the pupil how to learn, will really tell the teacher how to teach.

Someone will ask whether a book on the Senses can be called "Nature Study." Let Wordsworth answer in his lines on Tintern Abbey. The poet's definition of Nature would have been all-embracing :—

> ......"the light of setting suns,
> And the round ocean and the living air,
> And the blue sky, and in the mind of man."

The title of this little book is an echo of an earlier title: "The Five Gateways of Knowledge," by Prof. George Wilson, of Edinburgh, having been published by Messrs Macmillan & Co. in 1857; and by their goodwill we feel free to use this phrase again. Prof. Wilson makes his acknowledgments to "The Holy War" by John Bunyan. "This famous town of Mansoul had five gates...Ear-gate, Eye-gate, Mouth-gate, Nose-gate, and Feel-gate."

Prof. Wilson addressed his lectures to a philosophical audience; the present book is intended for use in Higher Primary and Lower Secondary Schools, that is to say, for pupils of about 12 to 15 years of age. For their sakes the wording is clear and simple.

In the last sixty years methods of education have become increasingly practical, hence abundant and varied practical exercises are suggested. Complete instructions are given for these exercises so that the book shall be intelligible in the absence of a teacher. The proposed practical work is not only what might be done by eager boys and girls on half holidays, but what can be done by every pupil in the course of ordinary school work. The pictorial illustrations chosen are aids to observation, not substitutes. Sufficient directions are given for the supply of necessary material.

HUGH RICHARDSON.

12, St Mary's, York.
1912.

CONTENTS

LIST OF ILLUSTRATIONS

Figures 12, 25, and 30 are from the *Encyclopaedia Britannica*; figs. 1, 2, 3, 4, 8, 10 *a*, 24, 29, and 35 are from photographs by Mr M. O. Dell; figs. 22 and 26 are reproduced from Dr C. S. Myers' *Experimental Psychology*, by permission of the author; figs. 23 and 41 are from Halliburton's *Handbook of Physiology*, by permission of Mr John Murray; the portrait of Sir Francis Galton (fig. 49) is reproduced from *Biometrika* (Vol. II) by permission of Mr E. G. Wheler; for fig. 14 thanks are due to Dr W. H. Drinkwater, of Wrexham, who kindly supplied the original radiograph.

CHAPTER I

THE MEANING OF OBSERVATION

Observation and hearsay—Illustrations of the meaning of the terms sensation, sense-organ, stimulus, sensibility

Apparatus and materials required.

Salt. Red, blue and yellow glasses (coloured gelatine may be substituted). Simple lenses. Sheet of brown paper. Watch. Ground glass. Ribbed glass. A thick glass block.

The meaning of observation.

This book deals with the methods and machinery of observation. Observation means obtaining information about an object by one's own senses.

EXERCISE 1. *To illustrate what is meant by observation.*

Make a list of all the facts you know concerning salt arranging them in two columns, (1) those facts which you can learn from your own personal experience by actually examining a pinch of salt, (2) those facts you know by hearsay or have read in a book.   Enter your results in your note-book thus :—

Facts about Salt.

From personal experience	From books or hearsay

Each of the facts mentioned in the left-hand column is a statement of the way in which salt affects the person examining it—in other words an " **observation**."   To use the technical term we may say that the salt gives us certain **sensations**, viz. of heaviness, whiteness, saline taste, and the like.

These sensations we receive by the body or some part of it.   Thus the sensation of saline taste reaches us by the tongue and that of whiteness by the eye.   A part of the body by which a sensation is received is called a **sense-organ**.   The eye and ear are sense-organs pure and simple.   The nose, tongue and palate, and the skin generally serve as sense-organs as well as having other uses.

EXERCISE 2.   *To illustrate the use of the sense-organs in making observations.*

Copy the list of observations made on salt and opposite to each write the name of the sense-organ used in making the observation.

Arrange the results thus :—

Observations on Salt.

Observation made	Sense-organ used

Anything like salt which acts upon a sense-organ and produces a sensation is called a **stimulus** which is the Latin word for a goad.

Not infrequently the same object, as a pinch of salt held in the hand, may act as a stimulus to more than one sense-organ at a time.

EXERCISE 3. *To illustrate the way in which stimulus, sense-organ, and sensation are connected.*

Make a table to show what sense-organs may be affected and what sensations may be produced by the following objects :—

> A flash of lightning.
>
> A glass of cold water.
>
> A peal of thunder.
>
> A pin.
>
> Violets.

Think of other objects which may act as stimuli and add them to the list. Arrange the results in three columns thus :—

Connection between Sense-organs, Sensations
and Stimuli.

Object acting as stimulus	Sense-organ affected	Nature of sensation produced

Methods of studying observation.

Suppose an educated man for the first time in his life could see a pair of spectacles.   He might study the curvature of the glasses, the shape of the frames, the construction of the hinges and so learn how the glasses were to be used.   On the other hand he might put on the spectacles and make a careful description of the change they produced in what he saw.

EXERCISE 4.   *To illustrate the two methods of studying observation.*

Examine the glasses provided.   Carefully describe them and then describe the appearance of objects when viewed through them.   Use the following kinds of glasses :—coloured glasses of various kinds, ground glass, ribbed glass, a thick glass block.

Make a rough ear trumpet out of brown paper ; describe exactly how it is made, and then describe the difference it produces in the sound of a watch ticking when the trumpet is held to the ear.

These exercises are intended to bring out the fact that in studying the methods of observation there are two quite different problems which can be studied concerning any sense-organ.   One of these is the question "how does the sense-organ work?" How is it e.g. that the eye enables us to see? On the other hand there is the question "what does it feel like to see?"

In the chapters that follow both of these problems will be touched upon with regard to each of the various sense-organs in turn. We shall see that different persons are not always equally affected by the same stimulus, and it will be shown that in many cases it is possible to compare their "powers of feeling." A person's power of feeling is called his **sensibility**. Thus it is pretty well known that some persons cannot hear a bat squeak, but that others can. Those who can hear such sounds are said to have a greater sensibility to shrill notes.

CHAPTER II

THE BRAIN, NERVES, AND SENSE-ORGANS

Arrangement of the brain and sense-organs in the skull—The backbone—The general structure of the brain of the rabbit—The spinal cord and its branches—The structure of nerve

Apparatus and materials required.

Sheep's skull divided down centre (or skull of rabbit, cat or dog). Backbone of rabbit, cat or dog. Piece of spinal cord from sheep or rabbit. Nerve from frog's leg or sheep's head. 2°/₀ salt solution. Mounted needles (large darning needles in corks). Formalin. Plasticine.

Bones and skeletons may be obtained ready prepared from Mr C. Baker, High Holborn, London, W.C., or

from Messrs Edward Gerrard & Sons, 61, *College Place, Camden Town, London, N.W. In this form however they are somewhat expensive.*

How to prepare a skull.

A good idea of the general arrangement of the nervous system and of the chief sense-organs is obtained by studying the skull and backbone of some animal, e.g. a sheep or a dog.   A sheep's head can be obtained

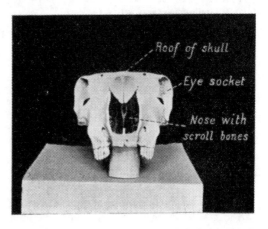

Fig. 1.   Front view of skull of sheep

at prices ranging upwards from fourpence, and cleaned dogs' skulls also can frequently be obtained or may be purchased from any of the regular dealers.   The most convenient way of preparing a skull for examination is to divide it in two halves lengthways.   If this is done before the skull is cleaned the process of cleaning is very much simplified.   Cleaning is best done by boiling till the flesh is quite soft when it can be scraped from the bones.   Soaking in water which contains a little chloride of lime and subsequent exposure to the sun will ensure the bones being white. Care should be taken in cleaning that the minute bones of the ear are not lost. They should be looked for and removed as soon as boiling has gone far enough to allow of this being done.

Examination of a sheep's skull.

Place the half-skull on the bench in front of you with the outer surface uppermost. Find the dome-like

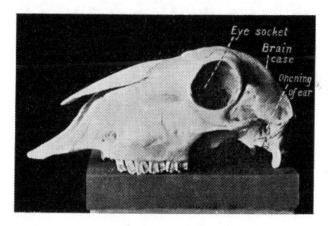

Fig. 2. Side view of skull of sheep

brain case which forms the hinder part of it. Notice the large hole at its hinder end ; through this during life runs the spinal cord or spinal marrow which passes right down the backbone.

Try and find out by studying the actual skull the places occupied by the eyes, the nose, the ears and compare with the figures given in this book. (Figs. 1—4.)

The **eye socket** is a circle of bone in which the eye-ball rests ; in the back of it near the brain is a small hole which puts the eye into communication with the brain case and through which during life the **optic nerve** or nerve of sight passes. Toward the front margin of the .eye socket is a small hole which passes through a plate of bone into the nose. Through this runs a small tube allowing the fluid with which the

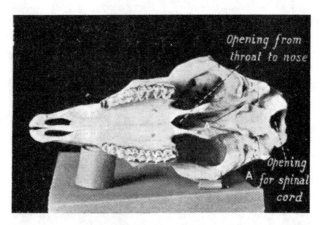

Fig. 3.   Lower surface of skull of sheep
(*The black threads at* A *pass through the ear passage*)

eye-ball is bathed to flow down into the nose. This is called the **tear-passage** or **lachrymal duct** from the Latin *lachryma* = a tear.

The **passage of the ear** is near the hinder end of the skull. A small hollow bone projects from the lower surface on each side. The position of this can be easily seen by looking for a large downwardly projecting peg at the back of the skull and then looking immediately in front of it. Each of these ear bones is like a very much flattened flask with the neck outwards. Across the opening of the neck during life stretches a thin skin, the **ear drum**, which thus shuts off the inner part of the ear from the outside. If a bristle is thrust into this hole in a forward direction, another opening can be found from this ear bone. Through this second opening passes a tube which opens from the ear into the throat near the back of the nose. (See Fig. 3.)

Inside this flask-shaped bone there is a chain of three very small bones which reach across from the ear drum to the more delicate "**internal ear**," where the actual hearing takes place. Unless special care has been taken in preparing the skull these bones are liable to get lost.

At the base of the skull is the **mouth** fringed on each side with a row of crushing teeth. Between these is a hard floor or shelf of bone, the roof of the mouth or the **hard palate**. Above this is the cavity of the nose which is roofed in its turn by two long and narrow bones called the **nasal bones**. To see these parts turn the half-skull round so that the inner half can be seen. Try and find out what everything you see is, first from the skull itself, then by referring to the figure and description given here. (Fig. 4.)

The cavity of the nose can now be seen from the inside and in it twisted plates of bone called **scroll bones**. These are covered over with a thin skin and kept moist during life but will of course be dry in the prepared skull.

Can you make out any connection between the cavity of the nose and the brain? At the hinder end of the nose is a plate of bone perforated with very numerous holes like a sieve. Through these run a multitude of fine nerves which are the **nerves of smell.**

At the hinder and lower end the nose opens into the throat. The presence of such a connection between nose and throat explains the fact that we can breathe either through mouth or nose and also the painful effects of a cough while drinking.

Only one other bone need be specially mentioned here. It is a paired solid bone best seen from the inside and lies toward the hinder end of the skull. It

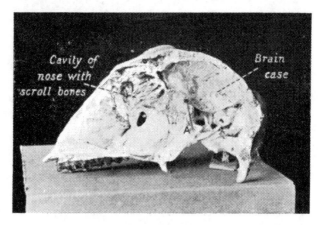

Fig 4.   Section of skull of sheep
(*The penholder at* A *is pushed through the aperture of the optic nerve*)

is, as it were, wedged into the wall of the brain case between two other bones; this is the bone which contains the **internal ear.** On its inner surface is a pit into which runs the nerve of hearing or **auditory nerve.**

EXERCISES.

1. Draw the skull of the sheep from the left side.

2. Draw the half-skull as seen from the inner surface.

3. Draw the skull as seen from directly in front.

4. Make a back view of the skull.

5. Carefully study and draw much larger than natural size the junction between two of the bones which roof in the brain case.

Examination of the backbone. (Rabbit's, cat's or dog's bones may be used.)

Examine a single bone or vertebra from the backbone. Select one, first, from the region immediately behind the ribs. Afterwards examine those of the neck and the region of the ribs and note any differences.

Each vertebra consists of two parts, (1) a **solid cylindrical block** which is placed toward the lower surface of the body.

(2) A more slender **arch-like part** pointing the opposite way and bearing various spines and projections to which during life muscles are attached.

Find out how the separate vertebrae are fixed together. From each end of the arch-like portion arise a pair of projections. Those at the hinder end fit in between those of the forward end of the next vertebra behind.

Fit two vertebrae together and observe the way in which the arches form a continuous tube and that small openings occur at the sides of this between them. A good idea of the shape of the cavity can be obtained in the following way. Place two vertebrae end to end and fill the tunnel formed by the arches with plasticine forced in from the ends. Then remove the bones and look at the rough cast which has been made.

EXERCISES.

1.   Make a sketch of a vertebra seen from the front, enlarged.

2.   Draw a vertebra seen from the side.

3.   Draw a vertebra from that part of the backbone which bears the ribs, as seen in various positions.

4.   Make sketches of a neck vertebra.

5.   Draw the rough cast made of the tunnel formed by the arches of the vertebrae.

6.   Draw vertebrae from any other animals obtainable.

The brain.

During the life of the animal the brain case is occupied by a soft pinkish-white mass of substance called the **brain**.   Figure 5 shows the brain of the rabbit as seen from the upper surface.   It is seen to be a roughly pear-shaped organ with the stalk end pointing forwards.   At the front end are two projections one on each side.

These are the parts specially concerned with the sense of smell ; they run down to the front end of the skull and are separated from the cavity of the nose only by the thin perforated plate of bone already described.

Behind these and forming the front two-thirds of the brain are two large pointed objects side by side, whose upper surface is rounded and smooth.   These are called the **hemispheres**.

Immediately behind them is a rather irregular mass covered over with folds and grooves in a complex way. It is roughly divided into three parts, one in the middle and one on either side, and may be conveniently named the **hind brain**.   Behind it the brain tapers very rapidly and passes into the spinal cord.

The spinal cord and nerves.

The **spinal cord** or spinal marrow is a long white cord which in a sheep is as thick as the little finger. It is seen as a soft pinkish-white string in the backbone when a carcass is divided lengthways by the butcher, hence the name "spinal marrow." A piece

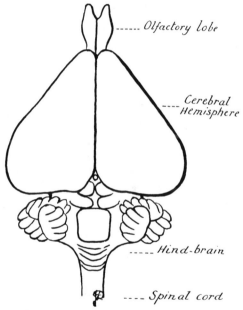

Fig. 5. Brain of rabbit seen from the back
(After *Wiedersheim*)

dissected from such a carcass will often show the form of the cord or if preferred a piece some two or three inches long can be removed from a rabbit.

The material used should first be hardened in formalin and then cut across at the end so as to give a clear idea of the form of the cord in section.

Place the piece of cord with the back surface uppermost and notice the groove which runs lengthways along the middle line.

Notice the nerves which arise in pairs from the sides of the cord. Each of these arises by two roots or two groups of roots one of them immediately behind the other. Is any other difference noticeable between the roots? Turn the cord over and examine the under surface. Note the groove running along the middle line as in the upper surface. Look at the cut end of the cord and see if any trace can be found of the small hole which runs along the tube from end to end. See also whether any distinction can be discovered between a gray H-shaped portion of the cord lying in the centre and a white portion lying toward the outside. Make careful sketches to illustrate what you see.

Each of the sense-organs is connected by fine threads, called nerves, with the brain. If the brain of a living animal is damaged it can no longer feel or act and may very probably die. If the nerve is divided no further sensation can be received from the particular sense-organ with which it is connected. A somewhat similar result follows when a nerve is tightly tied or compressed.

EXERCISE 5. *Observe the effect of sitting for* 20 *minutes with the legs crossed.*

The foot of the upper leg gradually "goes to sleep," i.e. ceases to feel, and in the end the power of movement may be lost. This is due to the compression of the nerve which supplies the skin and muscles of the foot.

The examination of nerve.

Obtain a short piece of one of the larger nerves of the frog. To obtain this most conveniently open a recently-killed frog from the lower side. Press all the organs to one side including the kidney of the other side. Behind this can now be seen a group of large nerves running backward parallel to the vertebral column. Any one of these will serve[1].

Suitable pieces of nerve can also be easily obtained from a sheep's head, which could also be used for the preparation of the skull as already explained, p. 6. A large nerve will be found crossing the face obliquely downwards passing from just beneath the eye toward the margin of the upper jaw. Wash the nerve carefully

Single Fibre   *Place where fatty sheath is absent*

Fig. 6. Longitudinal section of nerve
(*Magnified ; for explanation see p.* 16)

and examine it in a drop of salt water upon a glass plate. Notice the small fibres of which the nerve is made up. Try to separate these by means of a pair of mounted needles (two large darning needles stuck through corks will do). Notice that a sort of tough sheath holds the fibres together. If a microscope can be used place a short piece of the nerve teased out in salt solution beneath it and examine it under as high a power as can be used. See whether the form of the individual

[1] For details as to the dissection of the frog, examination of nerve, etc., *The Frog*, by A. Milnes Marshall, published by J. E. Cornish, may be consulted.

fibres can be made out. This is not very easily seen without a good deal of practice in handling a microscope, so a figure of a few nerve fibres is shown below. In the centre of each fibre is seen a **cylindrical cord**; round this lies a **fatty sheath** which in its turn is surrounded by a very thin **outer sheath** only to be seen with difficulty. The inner sheath is here and there interrupted by the outer sheath closing in so as to touch the central cylinder. One of these interruptions is shown in Fig. 6.

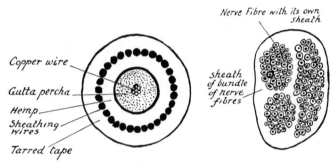

Fig. 7.   Transverse section of Atlantic Cable and nerve
(*Messages are carried by nerve fibres and by the copper wire*)

Nerves have often been compared to electric cables where the actual conducting part used is a small cord in the centre, all the rest being made up of various kinds of packing. The comparison is illustrated in Fig. 7 which shows transverse sections of nerve and cable side by side.

At the end furthest from the brain the nerve ends in a special kind of branched ending which serves as a receiving station for sensations. These receiving stations are so closely placed that there is practically no part of the skin where a point can be placed without being felt.

There are other nerves running out from the brain which do not end near the skin at all. Instead they end in a special manner in muscles. Their purpose is to convey messages from the brain to the muscles in which they end. The kind of messages conveyed by them are impulses that certain movements shall be carried out.

In the chapters that follow we shall study the experiences which are obtained by the action of various nerves and, later, the carrying out of actions.

CHAPTER III

THE SENSE OF TOUCH

The kinds of information supplied by touches—Position of touches—Size of touches—Pressure of touches—Comparison of touches—Weber's Law

Apparatus and materials required.

Miscellaneous objects, e.g. cork, pen, ink-pot, ball, etc. Dividers, knitting needle, corks, needles. Post-cards, fine silk thread, Bunsen burner and tripod with sand-bath. Test tubes, burette. Cartridge-cases, shot, saw-dust. Fine copper or aluminium wire. Aniline pencil.

Messrs Cussons of The Technical Works, Broughton, Manchester, supply flat brass slotted weights of values $\frac{1}{10}$ lb., $\frac{1}{20}$ lb., and $\frac{1}{100}$ lb. which would be convenient for some of these experiments.

The meaning of the term "touch."

A touch from any one object is really quite a complicated affair and may supply the person touched with quite a number of different kinds of information. This is illustrated by the following experiment.

EXERCISE 6. *To illustrate the different kinds of information which can be obtained from a single touch.*

Allow some object to rest for a few moments upon the palm of the hand. Without looking at the object answer the following questions : Is it hot or cold? heavy or light? Does it cover half the palm? or less? or more? What part of the palm exactly is it in contact with?

Repeat the experiment with a number of different objects, e.g. a pen, ink-pot, cricket-ball, a feather, a tumbler of warm water, etc. Write down one-word answers to each question for each object tried. Arrange the results in a table thus :—

	Pen	Ink-pot	Cricket-ball	—	—	—
Hot or Cold						
Heavy or Light						
Half-palm or more or less						
On						

Can any further observations be made about the objects? Is such observation really new or a combination of previous observations?

The results of this experiment show that any one touch may give the following kinds of information :—

(1) Information as to what part of the body is touched.

(2)   Information as to the size of the object.

(3)   Information as to the pressure that is exerted upon the skin by the object touched.

(4)   Information as to the temperature of the object touched.

Such information as is obtained in this way can be made more accurate by other means, e.g. if it is wished to judge temperature more accurately a thermometer is used, while pressures can be measured with weights and a balance.

We next go on to enquire with what degree of accuracy the sense of touch will work when unassisted by any apparatus.

Position.

The accuracy with which the position of a touch can be judged is most simply tested by discovering how far apart a pair of points must be in order to be separately distinguished.

EXERCISE 7.   *To see how far apart two simultaneous touches must be in order to be separately distinguished.*

Before beginning to experiment decide on the exact part of the body to be tested.   It is convenient to make a light mark with an aniline pencil upon it so as to be certain of the place, say the back of the hand.   A convenient apparatus for use is an ordinary pair of dividers such as is shown in Fig. 8.   The same figure shows a home-made piece of apparatus which could also be used for the same purpose.

Between the legs of the dividers at *A* is shown a small narrow wooden wedge which is a useful means of slowly separating the points.   In the alternative apparatus to the right *B* is a knitting needle on which

2—2

two small corks *C* are slipped.   The letter *D* is at the
heads of two needles whose points are driven into the
corks.   By sliding the corks along the knitting needle
the heads of the needles at *D* can be separated by any
interval required.

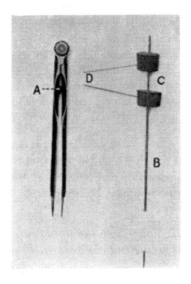

Fig. 8.   Apparatus for testing accuracy of
judgment of position by touch
(*At* A *is a small wedge to separate the
legs of the dividers*)

In making the test the person tested, who for convenience is often referred to as the **subject** of the
experiment, should know what is being done but should
not watch the experiment.   Take the pair of dividers
or the needles and arrange them so that the two points
can be quite easily distinguished as two.   Touch the skin
at the place to be tested and repeat the experiment,

gradually bringing the points nearer and nearer together until a difficulty is found in distinguishing between one touch or two. In this way a distance can be found which is roughly speaking the answer to the problem suggested at the head of this exercise.

To obtain a more accurate result account must be taken of the fact that the answers given by the subject will not always be the same for the same distance. Thus in one experiment he may say "two" when the points are 5 mm. apart, yet on a second trial he may answer "one" for the same distance. It is usual to get over this difficulty by taking that distance at which eight out of ten answers are given correctly. Having found a rough result as described above, the apparatus is next set at a distance just great enough to enable the two points to be separately distinguished. Ten trials should now be made with the points separated by this distance and among them arranged in irregular order ten cases in which only one touch was used.

The answers given should be recorded in a table like the following :—

How far apart											No. of times two pts. correctly recognised
11 mm.	⎰ No. of pts. used	1	2	2	1	1	1	2	2	etc.	—
	⎱ Answer given	1	2	2	1	1	1	2	2	etc.	10
10 mm.	⎰ No. of pts. used										
	⎱ Answer given										
9 mm.	⎰ No. of pts. used										
	⎱ Answer given										
etc.					etc.						etc.

Test in this way the following parts of the body :
(1) centre of palm, (2) centre of each finger tip, (3) back
of forearm two inches above wrist, (4) back of hand,
(5) knuckles when fingers are bent, (6) the same when
fingers are extended, (7) any other part of the body
easily accessible.

" Size " of touches.

It has been shown in a preliminary experiment that
sensations of touch can be distinguished according to
size. A large burn feels different from a small one even
though both may have been produced by objects of
equal temperature. In our ordinary experience we do
not distinguish this feeling of size from the local quality
of the sensation, i.e. the sensation occupies a space and
this space is definitely felt by us to be on some special
part of our bodies. If a glove is put on the left hand
and another put on the right, two sensations are produced which are about the same, as far as size is concerned, but totally different as to place.

To test the accuracy with which the size of an object
can be judged by the sensation it produces, a method
similar to that used in the last experiment may be
employed. We may discover e.g. what is the shortest
line which can be distinguished as a line and is not
merely felt as a blunt point.

EXERCISE 8. *To see how short a line can be distinguished from a point by touch.*

Cut out a series of cardboard oblongs of any convenient length to handle, e.g. 10 cm., and of various
widths ranging from 1 mm. to 20 mm. (Fig. 9.)
The strips may be very easily cut out by pasting

a sheet of squared millimetre paper upon the card and using a sharp knife.

Get your partner to touch you with each of these lines in succession until one is reached which is just recognisable as a line.

Repeat the exercise but start with a line easily felt as such and work downwards until a line is reached which seems exactly like a blunt point. The mean of these two results may be used as the shortest line recognisable as such.

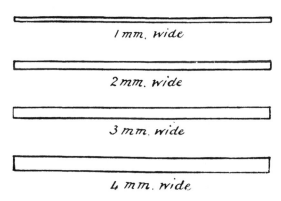

Fig. 9.   Method of making card edges of various lengths

This method is extremely rough but it would be possible without difficulty to modify the method of Exercise 7, p. 19 to suit this one. The working out of this is left to the pupil as a problem.

Pressure.

We next proceed to make similar experiments with regard to sensations of pressure.

EXERCISE 9. *To see what is the smallest weight which can be detected by touch.*

Take an ordinary thin post-card and find its weight, or better, cut out a square of thin card whose side is 10 cm. and weigh that. Find the area of the card in use and from the results calculate the weight of 1 sq. cm. of it. Next cut out a series of strips of the same card 1 cm. wide and of various lengths, e.g. 2 cm., 3 cm. and so on up to 10 cm. The areas of these strips should next be written upon them and their weights calculated.

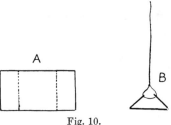

Fig. 10.
Weight for finding least perceptible weight
A = *card*
2 *sq. cm. enclosed by dotted lines = folds*
B = *card weight ready to use*

They can then be attached to very fine threads of silk about 6″ long. The attachment is best secured by folding up the edges of the card so that it can rest on a surface 1 cm. square and then passing the silk through the turned up edges. Fig. 10 shows a diagram of such a weight and Fig. 10 *a* shows the weight in use.

The apparatus is now ready for use. The person to be tested closes his eyes or turns away his head and the experimenter *gently* lowers the cardboard weights in turn on to the hand or the part of the body to be tested, until a distinct weight is felt.

The person making the experiment records the first weight which is felt by the subject. As in previous experiments a difficulty may arise from the fact that the subject will sometimes feel a lighter weight than at other times and, as before, this may be overcome by selecting that weight which is felt eight times in ten trials.

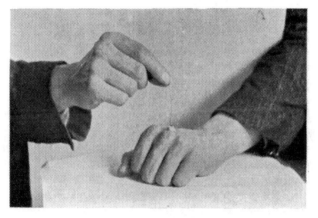

Fig. 10 *a*.
Cardboard weight as used for testing least perceptible weight

It is well to check the calculated weights of the cards by actual weighing as cardboard is apt to vary somewhat in weight.

It is also important to note that it is quite possible to feel an object as hot or cold without being aware of any pressure exerted by it. To avoid the difficulty which this causes it is convenient to have at hand a Bunsen burner turned low under an asbestos card or sand-bath. The weights can then be kept at about body temperature.

Use these weights to test the following parts of the

body : centre of palm, finger tips, ball of thumb, back of hand, back of fore-arm, cheek, forehead.

Do the small hairs with which the skin is covered affect the result? What difference if any is produced by shaving them off with a razor or a sharp knife?

What effect is produced if during the experiment the mind is concentrated on something else, e.g. a stiff piece of multiplication?

Comparison of sensations.

So far we have tested sensitiveness by finding out how small a stimulus can be just detected at all. We may measure equally well the amount by which a stimulus once felt must be increased in order that any difference may be noticed, e.g. we found in Ex. 8 that until the points of the dividers were a certain distance apart they were not felt as two points at all. How much must the distance between them be increased in order that they may be felt as distinctly wider apart? Or again (Ex. 9), it has been seen that until a line is a certain length it is not felt as a line at all; by how much must its length be increased when once it has been felt in order to seem definitely longer? We should in this case be measuring the power of distinguishing between objects or power of **discrimination**.

It may be remarked by way of illustration that supposing it is desired to test the accuracy of a balance the same kind of experiment might be made. Either very minute weights could be used and that one determined which would just cause the balance to turn visibly; or a certain "load" might be placed in the pans and one side gradually increased until it had been found out by what fraction of itself a weight must be increased for the balance to detect the difference. [Really the plan more often adopted is to put a known load, say 100 grms., on each pan with an extra ·01 grm. on one of the pans and measure on a divided scale the corresponding deflection of the pointer.]

So in experimenting on the sensation of pressure we have first determined how large a weight must be to be felt at all; we next attempt to discover how much a weight must be increased to feel distinctly heavier.

EXERCISE 10. *Find out by what fraction of itself a weight must be increased in order that the difference may be noticed.*

Take two test-tubes and seccotine on to the bottom of each 1 square centimetre of card. When the card has become firmly fixed, find the weights of the two test-tubes and counterpoise them exactly with shot. Fill a burette with water and read the level.

Place the two test-tubes in succession upon the part to be tested and then gradually run into one of them a few c.c. of water from the burette. Again place them in turn upon the area of the skin to be tested and see if the increase in weight is noticed. Go on adding water to one of the tubes until there is a noticeable increase.

A preliminary trial will show about how much water must be added. Successive trials can then be made in which the water can be added fairly rapidly at first and the point where the difference is first perceived can then be approached more gradually.

A more exact method of carrying out the same experiment.

A more exact method of carrying out this experiment consists in using a set of weights similar in appearance but varying in weight from say 15 grms. to 25 grms. These can be made in the manner suggested by Galton in *Inquiries into Human Faculty*[1]. It is also easy to make them of test-tubes with cards stuck on the rounded ends. These can then be made up to the necessary weights with shot. It is clearly a useful plan to start by finding the average weight of one shot. When such a series of weights has been prepared, four different determinations can be made with them.

Starting with a certain standard, let us say 20 grms., the standard may be compared with 21 grms., 22 grms., etc., until a weight is reached which is recognised as heavier.

On the other hand the 20 grm. standard might be compared first with a clearly different weight ; say 25 grms. or some other number shown to be large enough by a rough preliminary test, and then with a gradually decreasing series 24 grms., 23 grms., 22 grms., etc., until a weight is reached which the person experimented upon fails to detect as different.

In just the same way the standard weight of 20 grms. might be compared with weights less than itself, and again in two ways first starting with two weights which seem the same and afterwards with two weights which seem different.

In this way four values would be obtained for the just observable difference between 20 grms. and any other weight. The mean of these four values is usually taken as the just observable difference.

The result is still somewhat rough as we have taken no means to ensure that the order in which the two weights are tried in each comparison does not affect the result. To ensure this it would be necessary to

[1] "Everyman's Library" Edition, p. 250.

repeat every trial twice, first taking the standard weight first and then the other. The following table shows all the series of comparisons which could be made.

Repeat this experiment with as much accuracy as time will allow and taking different starting points, e.g. 20 grms., 40 grms., 50 grms., etc.

A somewhat different method of experimenting with the same weights by means of the muscular sense or sense of resistance to movement will be described in Chapter VI.

Standard = 20 grms.

Comparison with lighter weights	Comparison with heavier weights
Standard first.	*Standard first.*
Going away from standard, 20 and 19, 20 and 18, 20 and 17, etc.	Going away from standard, 20 and 21, 20 and 22, 20 and 23, etc.
Going towards standard, 20 and 15, 20 and 16, 20 and 17, etc.	Going towards standard, 20 and 25, 20 and 24, 20 and 23, etc.
Standard second.	*Standard second.*
Going away from standard, 19 and 20, 18 and 20, 17 and 20, etc.	Going away from standard, 21 and 20, 22 and 20, 23 and 20, etc.
Going towards standard, 15 and 20, 16 and 20, 17 and 20, etc.	Going towards standard, 25 and 20, 24 and 20, 23 and 20, etc.

Weber's Law.

Carefully arranged experiments of this sort have shown that the following fact is *broadly speaking* true. **If two weights are taken which are just distinguishable, the difference between them will be a certain definite fraction of the smaller weight no matter what the smaller weight is.**

To take an actual example ; if a weight of 21 grms. rested on the finger tip is just noticeably different from a weight of 20 grms., then any standard weight will have to be increased by $\frac{1}{20}$ of itself before the difference is felt. If for instance a 40 grm. weight had been the starting point, the increase necessary would have been $\frac{1}{20}$ of 40 grms., i.e. 2 grms., or in other words a 42 grm. weight would be just distinguishably heavier than a 40 grm. weight.

The following exercise may help to make the point clear.

EXERCISE 11. *Given that any weight must be increased by 1/n of itself to be felt as distinctly heavier, what weight will be just noticeably heavier than x grms. ?*

The law which has here been described at some length for sensations of weight is named after its discoverer **Weber's Law**. It applies not only to sensations of weight, but also to very many sensations such as light, loudness of sounds and many others.

Its effects are very numerous in everyday life. A rich man gives 1s. and a poor man 2d. to a porter who does the same service for each. The reason is in part that they each give a just noticeable fraction of the contents of their pockets.

EXERCISES.

1. What difficulties are met with in tying a piece of cotton ? Why is it easier to tie a piece of rope ?

2. Repeat Ex. 6 but instead of bringing down both points at the same moment bring one down a fraction of a second before the other.

3. Given that a line must be increased by 1/n of its own length to seem longer, how long must a line x cm. long be made to be felt as distinctly longer ?

4. Place a test-tube with card on the end as used in Ex. 9, p. 24 on the back of the hand and run water in slowly from a burette. How much must be added before the test-tube feels heavier ? Compare with the result of Ex. 9.

5. Place 100 grms. on each pan of a chemical balance. Then add cautiously 5, 10, 15, etc., mm. of very fine aluminium or copper wire until the pointer swings over by 1 scale division. Repeat the experiment with a load of 200 grms. on each pan and also with lighter loads. Do you find any connection between the load and the length of wire which must be added to make the balance turn ?

CHAPTER IV

HEAT, COLD AND PAIN

The indifference point—Circumstances which affect the indifference point—Heat and cold spots—An illustration of pain

Apparatus and materials required.

Thermometers, beakers, tripods, with gauze or sand-bath, Bunsen burners, glass rod.

Information received by the skin about temperature.

It has already been noticed (p. 19) that any object touched is felt to be either hot, cold, or indifferent.

EXERCISE 12. *To find at what temperature water feels neither hot nor cold.*

Place a beaker of water on a tripod over a Bunsen burner turned very low and warm very slowly. Stir every few minutes with a thermometer. At intervals, say every five minutes, push the forefinger into the water up to the knuckle and note (1) the temperature and (2) whether the water feels hot, cold, or neither. It is convenient to do the experiment two or three times over, in the first case passing fairly rapidly up the scale and in the second proceeding more slowly when approaching the point where the water feels indifferent.

This experiment illustrates the fact that there is a certain temperature at which a substance feels neither hot nor cold to the skin. This is sometimes called the **indifference point**. Our next two experiments will illustrate two conditions under which the indifference point may change.

EXERCISE 13. *Find the indifference point when the part tested is kept cold between the times of testing.*

Repeat the last exercise but keep the finger between times in a stream of cold water or else immerse for a few minutes in ice cold water just before each test.

EXERCISE 14. *Find out what effect if any is produced on the indifference point by altering the surface of the skin acted upon.*

Repeat Exercise 12 but in each case immerse two fingers or, if the vessel is large enough, the whole hand in making each test.

The results should show that it is quite possible to shift the indifference point from its usual position by previously exposing the part tested to a different temperature and further that the indifference point depends, at least in part, upon the area of the skin affected.

To obtain more exact information on the subject of the accuracy with which temperatures can be judged by hand the same kind of method could be used as has been suggested in the case of other sensations. Thus

we might start with a body of indifferent temperature
and see how far from that we must go before we get
any feeling of temperature at all, hot or cold as the case
may be. This is like finding the least noticeable weight.
On the other hand we might start with an object which
felt hot and find out how much its temperature must be

Fig. 11.
Glass rod, as used for showing
heat and cold spots
(*The beaker contains hot or cold
water as required*)

increased in order that it should feel distinctly warmer.
This would be like finding the least noticeable increase
in a given weight.

To carry out such experiments carefully graduated
thermometers are required and the methods will not be
dealt with here. For fuller information concerning them

reference should be made to the works cited in the appendix.

Heat and cold spots.

EXERCISE 15. *To show that certain parts of the skin are specially sensitive to heat and cold.*

Take several pieces of glass rod and draw each out to a blunt point. (See Fig. 11.)

Allow some of them to stand in hot water, others in the coldest obtainable.

Take a rod from the hot water, wipe it dry quickly and draw the point slowly over the back of the hand.

Repeat with one of the cold rods.

Do you not find that the hot and cold points of the rods are more easily felt as hot or cold in certain positions than in others?

There are in fact small spots on the skin which are specially sensitive to cold and others specially sensitive to heat. These are sometimes referred to as **cold spots** and **heat spots**.

See if you can map out a cold spot by drawing the cold rod across it again and again along parallel lines very near together.

Pain.

Such a sensation as that of warmth is pleasant. So also sensations of pressure if not too strong are by no means disagreeable. In either case however if the sensation is made very intense it takes on the special character of pain.

EXERCISE 16. *How hot must water be in order to be felt as painful.*

Take a convenient quantity of water in a beaker or

a small saucepan and raise the temperature slowly as in Exercise 12. At intervals stir with a thermometer, read the temperature, and feel with the forefinger, immersing it just as far as the crease of the knuckle each time.

Record the temperature at which the water just begins to feel painful.

EXERCISES.

1. Make a careful enlarged drawing of a clinical thermometer twice natural size. What is the temperature marked by the arrow? At what temperature would a clinical thermometer be likely to burst? Why does a clinical thermometer have to be shaken before being used?

2. Find the temperature of your mouth. If you cannot do this accurately try to take the temperature of your neighbour's mouth, both Centigrade and Fahrenheit. This is called blood heat. Use a clinical thermometer if you can get one.

3. Take the temperature of a hot bath or basin of hot water and record:

	° C.	° F.
Unpleasantly chilly		
Not quite hot enough		
Hot enough		
Not too hot		
Rather too hot		

How do these temperatures compare with blood heat, 98·4° F.?

4. Take two thermometers; immerse the bulb of one in hot water and rest the other on the back of the hand. After a few minutes remove the one from the hot water and place it alongside the other on the back of the hand. Say when the two bulbs feel equally warm. Note the actual difference in temperature between them as shown by the thermometer scales.

CHAPTER V

THE MACHINERY OF MOVEMENT

Movement and touch—Bones and possible movements—Hinge joints and ball-and-socket joints—Muscles and how they produce movements

Apparatus and materials required.

Miscellaneous objects, e.g. ball, sponge, stick of sealing wax, etc. Skeleton of rabbit. Sheets of drawing paper 24″ × 6″, cardboard, elastic, paper fasteners. Dead rabbit.

Movement and touch.

Experiments have been described earlier in this book showing how certain kinds of information can be obtained about objects which are simply allowed to rest upon the skin. If however further information is required about any thing than can be obtained by merely touching it, it is next handled, i.e. the fingers or the hand are passed over the surface of the object, it is lifted, pinched and so on. By treating the object in this way further information is obtained about its weight and size and also one learns whether it is hard or soft.

EXERCISE 17. *To compare the information obtained by handling an object with that obtained by merely touching it.*

Handle, with the eyes closed, a stick of sealing wax, a lump of putty, a large cork, a block of wood, etc. In each case try first the effect of resting the object for a few moments on the hand and then the effect of moving it, squeezing it, etc. Tabulate your results to show how far the information given by the second

Fig. 12.   Bones of human arm

method (*a*) supports and (*b*) adds to that obtained by the first.

TABLE

Object	Result of touching	Result of handling	Additional information gained by handling

The means by which parts of the body move.

Examine the skeleton of a rabbit, in particular that of one of the limbs. Examine the fore-limb first and afterwards compare carefully with the hind-limb. (See Fig. 13.) Also compare the human arm, Fig. 12.

Notice that the fore-limb is attached to a large blade-shaped bone, the **shoulder-blade**, which has a ridge running along one side of it. To this the bone of the upper arm is fixed by a **ball-and-socket joint.** Find out by experimenting on the bones of a rabbit in how many different directions the upper arm can be moved on the shoulder-blade.

Next examine the bones of the fore-arm and find out how they are joined together and also how they are joined to the bone of the upper arm. The name **hinge joint** is sometimes given to a joint of this kind. When the upper arm is kept fixed, in how many different directions can the fore-arm be moved upon it ?

Turn finally to the bones which make up the fore-foot or "hand" of the rabbit. The general plan on which the bones of the fore-limb and hand of any land animal are arranged is the same. The easiest way to understand the bones of the rabbit's wrist and hand is to compare them with the general plan. Nearest the arm we usually find a row of three small squarish bones.

Then follows a single central bone opposite the middle of the row ; then a second row of small bones like the first but five in number. These bones make up the wrist. Following them we have a row of five longish bones from the outer end of each of which a finger springs. The long bones are all covered by the same skin but each finger has its own sheath of flesh and skin.

The fingers themselves are made up of differing numbers of shortish bones which are sometimes named **phalanges.** A diagram of this general plan is given below.

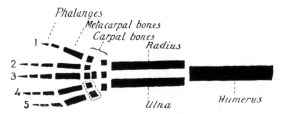

Fig. 13.   Diagram of bones of a fore-limb
(The two bones which are fused in the rabbit are surrounded by a dotted line)

The fore-limb of the rabbit is exactly like this, but two of the small bones in the row farthest from the arm fuse into one. The two which fuse are the two on the thumb side of the wrist. The thumb of a rabbit is not easily recognisable in itself but can be found by remembering that a rabbit's fore-feet are planted palm downwards and the thumbs are accordingly toward the middle side.

Make a drawing of the bones of the fore-foot of the rabbit together with the bones of the limb which carry it.

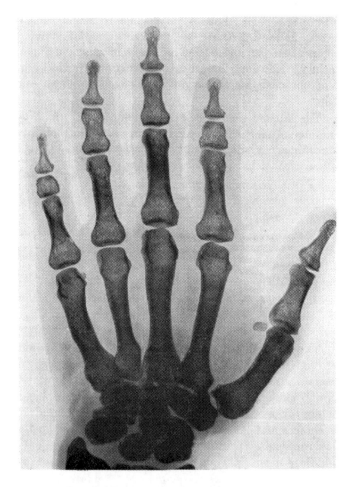

Fig. 14. Radiograph of human hand and wrist

EXERCISES.

1. Make a drawing of the bones of a rabbit's hind-limb and foot and compare with those of the fore-limb.

2. Try to make out by feeling through the skin how the bones of your own hand and arm compare with those of the rabbit.

3. Fig. 14 is a radiograph of the bones of a human hand and wrist. Try by holding your hand in front of a very strong light, say a magic lantern, whether you can find the corresponding bones in your own hand.

Compare the result with the bones of the rabbit.

4. If a living or recently killed rabbit or cat can be obtained, examine the foot carefully to see in what directions the various joints will bend on one another.

5. Examine the skeleton of a rabbit and find out all the places where there are (*a*) ball-and-socket joints, (*b*) hinge joints. Do you find any joints which seem immovable ?

Possible movements of a human arm.

Holding your shoulder-blade still, find out in how many different directions the upper arm can be moved. The easiest way to do this is to make a mark on a wall level with the shoulder, and standing sideways to it point the elbow at it. Then try whether the elbow can be pointed to a spot in front, above, below and behind the mark in turn. One other possibility of movement remains ; point the elbow at the spot and see if it can be turned round on itself still keeping it pointing in the same direction.

Try a similar experiment with the fore-arm pointing the wrist to a fixed mark. Use exactly the same method but take special care not to move the upper arm. This can easily be managed by holding it to the side of the body with the other hand ; any involuntary movement can then be at once checked.

The wrist is not a simple joint and the individual

bones are many and not arranged in a simple manner. The net result of their movements on one another may be made out by the same methods as have been used above. The wrist acts like a two-jointed hinge, i.e. it will bend so as to carry the hand towards the palm or towards the thumb or of course in the opposite directions.

If the above experiments have been carefully worked through it will have been seen that both the upper arm and the fore-arm can be twisted round without any change in their direction. How is this turning accomplished in the fore-arm? Turn the hand back and forwards several times and at the same time feel the bones of the fore-arm at different points with the other hand to see how they move on one another. Previous examination of the bones of the rabbit will have shown that in the lower arm there are two bones one ending on the thumb side and one on the little finger side of the arm. When the hand is twisted the lower end of the bone on the thumb side rolls over the other one. The upper ends of the bones do not change their positions, so that when the palm of the hand is upwards they are parallel but when it is downwards they cross obliquely.

The field of touch.

If a man had a stiff arm fixed by a single hinge joint like a doll's arm he would only be able to touch points in a certain particular position, namely those which lay upon the circumference or part of the circumference of a circle whose centre was the hinge and whose radius was the length of the stiff arm. If instead of the hinge joint his stiff arm was fixed by a ball-and-socket joint, the points he could touch would lie upon the surface of a hemisphere. If we imagine this stiff arm to have a second joint hinged to the end of it and then suppose the two to move independently, the end of the second joint would be able to reach an immense variety of points since every point on the surface of the hemisphere made by the end of the first joint would serve as a centre for movements of the second.

Now the human arm and hand, as we have already seen, consist of many joints, some able to move in one direction and some in another so that, as a result, there is an almost infinite number of points which can be touched. In fact there is a space reaching nearly three feet forwards and on each side of the shoulders within which any point can be touched by one hand or the other.

So far we have been dealing merely with possibilities of movement without considering what it is which supplies the motive power.

EXERCISE 18. *Examination of the muscles of the fore-arm.*

Examine your left fore-arm and hand with the palm surface upwards. Lay the right hand upon the left fore-arm and slowly clench the left fist and bend it upwards ; notice the change in shape of the fleshy part of the arm and the appearance of at least one thick cord in the wrist called a **leader.** By feeling with the tips of the fingers in the wrist find out if any other cords can be found. Make a sketch, natural size, of your wrist and hand to show the features which are revealed by the movement of the hand in this way.

Turn your arm so that the little finger and edge of the fore-arm rest upon the table. Place beneath them a sheet of drawing paper. Trace the outline of the arm (*a*) with the hand extended, (*b*) with the hand closed and bent as before.

If possible make experiments and drawings showing whether similar changes take place in the upper arm when the fore-arm is moved at the elbow.

Find out what changes take place in the calf of the leg when the foot is moved at the ankle.

Dissection would show that what has so far been called the fleshy part of the arm is in reality made up of elastic bundles called **muscles.** These are generally long and narrow and are fixed at their ends to the bones which have to be moved. The actual connection is made by means of tough cords called **sinews.** Sinews have already been seen in the wrist, where they are unusually long and receive the special name of leaders. When a movement takes place the muscle changes its shape, becoming shorter and broader. As a result the two parts to which the ends of the sinews are attached are brought nearer together, and although this actual amount of shortening of the muscle may be very small the movement of the parts to which it is fixed may be considerable. This will be easily understood by means of a model.

EXERCISE 19. *To make a model to show the way in which the muscles act.*

Take two strips of stiff card about 20 ins. long and 2 ins. wide. Fix them together by means of a paper clip passing through the end of one and about 3 ins. from the end of the other. These are to represent the bones of a simple hinge joint. To represent the muscles take two pieces of elastic and fix them as shown by the dotted lines in Fig. 15. An easy way to fix them is to

make holes through the cards and fix hooks of copper
wire to the ends of the elastic. The elastic should be
short enough to be slightly on the stretch in whatever
position the cards are. To show how the joint is moved
by the muscles, start with the cards and elastic arranged
as in the diagram. Then unhook one piece of elastic.
The other piece will then contract and a movement of
the cards on one another will be the result.

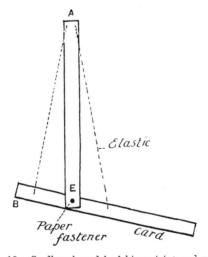

Fig. 15.   Cardboard model of hinge joint and muscles

The exact nature of the movement which occurs
when a muscle-change of this sort takes place in the
body depends entirely upon which points are free to
move. Thus if we imagine that our model represents
the elbow joint, then *A* will be the shoulder, *B* the
elbow, and *D* the hand. If *AB* contracts *D* will move
downwards. But suppose *D* is unable to move. Suppose
for instance that the owner of the arm is sitting in an

armchair and places the hands on the arms of it. If under these conditions AB contracts the only result that can follow is that E will be raised and the whole of the upper arm and body with it.

This model will illustrate very well the motions of the elbow, while if B stands for the heel, D the toes, and AE the lower part of the leg, the same will serve to illustrate the action of the ankle.

EXERCISES.

In the following exercises the model should be used and, if possible, the actual bones compared stage by stage with it.

Use the model to illustrate the following motions of the arm—

1. Using a hammer or chopping wood.
2. Lifting a cup of tea.
3. Pulling up on a horizontal bar.
4. Pushing in a drawing pin.

Also illustrate the following motions of the ankle—

5. Raising a weight on the toe.
6. Rising on tip-toe.
7. Tapping the floor with the toe while the heel rests on the ground.

Those who know anything of elementary mechanics will see that in the arrangement of bones and muscles we really have examples of the apparatus known as the lever. Just as in mechanics it is possible to distinguish three classes of levers according to the relative positions of the power applied, the point on which the lever turns and the weight moved, so in the body the same three classes are found. This point is referred to here merely for the benefit of any who happen to know something of mechanics ; a full treatment of it is beyond the scope of this book but may be found in Huxley's *Elementary Physiology*, p. 312.

CHAPTER VI

THE EXPERIENCE OF MOVEMENT

What we learn by moving—Smallest possible movements—The sizes of movements—How the sizes of movements are compared—The method of "average error"—Resistance overcome by a movement—Galton's weights—Fatigue—Fatigue of the senses—A caution to experimenters

Apparatus and materials required.

Retort-stand. Weights. Drawing-board. Test-tube large enough to slide over finger. Strip of wash-leather. Tacks or drawing-pins. Post-cards. Millimetre paper. Hanging spring balance. Spring balance made with elastic, or household scales. Two feet of thin board with a hole in the centre. Large sheet of patterned wall-paper. A bunch of strongly scented flowers, e.g. wallflowers, violets, pinks, lilies or roses, according to season. Dynamometer. Chip boxes. Cussons' series of weights $\frac{1}{10}$ lb., $\frac{1}{20}$ lb., $\frac{1}{100}$ lb.

Kinds of information obtained from movements.

A single movement, like a single touch, gives us impressions of many kinds. Without enabling us to distinguish every one of these different kinds of impression with absolute accuracy the following experiment illustrates two of the principal kinds.

EXERCISE 20. *To illustrate two of the principal kinds of impression produced by muscular movement.*

Get your partner to hold his hands about one foot apart horizontally and closing the eyes move your finger from one hand to the other. Repeat the experiment but in the second case increase the distance through which the finger travels. The feelings obtained by making these two movements are not the same, for in one case the finger is felt to have moved further than in the other. It follows that one kind of information supplied by a muscular movement is that of the **distance moved.**

Now repeat the experiment in another way. Take a retort-stand and fix a clamp or a ring 20 or 25 cm. above the base. Raise the fingers from base to ring several times, in each case carrying a weight. Begin with 10 grms. and try in succession 20 grms., 30 grms., 50 grms., 100 grms., 500 grms., 1000 grms. Notice what difference there is in the various sensations produced. Here we have sensations of movement which are the same as far as distance travelled is concerned, but which differ in the amount of energy required for carrying them out and in the **feeling of resistance** which is overcome.

The sizes of movements: active and passive.

Proceeding on the same plan as has already been followed we now go on to measure the smallest movement which can be felt at all.

So far the only movements which have been considered have been those which are produced intentionally. These are known as **active movements.** Another kind of movement is also possible. It is the sort which takes place when a fixed-wheel (i.e. not free-wheel) bicycle runs away down a hill. The legs of the rider are moved up and down involuntarily by the pedals. Such movements are called **passive movements.**

EXERCISE 21. *To measure the smallest angle through which a joint can be moved voluntarily.*

The size of the angle through which a joint turns is not the only means of measuring the amount of movement, but in this case it happens to be a simple and convenient method.

Fix a test-tube over the end of the forefinger like a long thimble reaching down to the knuckle.. Carefully

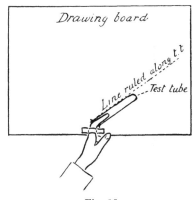

Fig. 16.

Apparatus for measuring the smallest angle
through which a joint can be turned

fix it in position by means of small wedges of paper so that the finger cannot move without moving the test-tube also. Fix the first joint of the finger to a drawing board by means of a strip of wash-leather and drawing pins. (See Fig. 16.)

Lay a ruler alongside the test-tube and rule a line along it with a sharp pencil to indicate the direction in which the finger is pointing as shown by the edge of the test-tube. Bend the finger by as small an amount as possible, again lay the ruler alongside the test-tube and rule a second line. The lines must then be produced till they meet and the angle between them is the angle through which the finger has been moved.

Repeat the experiment several times and take the mean of your results.

Use this method to compare the accuracy with which the fingers of the right and left hands can be moved.

Repeat the experiments, but instead of moving actively, allow the finger to be moved passively by some one else.

The method of average error.

The measurements just made correspond to measurements of the just observable weight, the shortest card-edge perceptible as a line, etc.   Following the same method we have used before we naturally next try to measure the power of distinguishing between movements of different sizes.   To do this a method which has not been used before is convenient.   It consists of making a series of movements all intended to be of the same size and then measuring the average error.

Testing the keenness of observation has been compared earlier in this book to testing a balance.   The same illustration may help to make clear what is meant by the **method of average error.**

If it were desired to test whether a certain balance could distinguish between one weight and another, it would be possible to do so by placing equal weights on the two scale pans and then gradually adding to one until the pointer of the balance showed a visible movement.   This minimum weight to make a just visible

movement of the balance will largely depend upon how
sharp or how blunt the knife edges are. If the knife
edges are rusty, perhaps a centigram will not make the
balance stir at all. If the edges are really sharp a tenth
of a milligram may be enough to move the pointer a
little, although so little that a magnifying glass may be
required to see the movement.

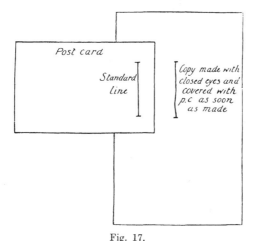

Fig. 17.

Materials for testing the accuracy with which the
amount of movement can be judged

As an alternative we might keep a fixed weight on
one pan and weigh out on the other several amounts of
sand or shot which the balance thus indicated as just
equal to the weight. According to this balance all these
weights of sand or shot would be equal to one another.
Now let them all be weighed on a much more accurate
balance. If the balance which was being tested was
really perfectly correct they *would* all actually weigh.

the same. If it was not very delicate, some would be too heavy and others too light, while some might happen to be correct notwithstanding. The more inaccurate the first balance the greater would be the errors, or differences from the correct result, as obtained on the second balance. The average error will therefore give a good idea of the accuracy of the first balance. It should be noted that we are not considering such inaccuracies as would cause a certain fixed error every time, as would be the case if the pans were unequal in weight, but simply the certainty with which the first balance will distinguish between one weight and another.

EXERCISE 22. *To find the accuracy with which a set of muscles can be made to move through a fixed distance using the method of average error.*

Rule a line 5 cm. long on a card and place it in front of you on a piece of paper in such a position that the line is vertical. (Fig. 17.) Take a sharp-pointed pencil and draw it twice down the line from end to end. Close the eyes and make a vertical mark on the paper which feels to be the same length as the line, cover with the card and repeat. Do this ten times but do not examine the results until all are completed. Then measure each of them as accurately as possible. Calculate the mean of the ten results.

Here is the result of an actual experiment.

No. of trial	1	2	3	4	5	6	7	8	9	10
Length of line drawn	4·15	4·34	4·20	4·56	4·14	4·27	4·28	4·67	4·24	4·25

The mean = 4·31 cm.

We next proceed to find how much each result differs from the **mean.** The first is ·16 cm. smaller. The second ·03 cm. larger, etc.   In this way we obtain the following table

No. of trial	1	2	3	4	5	6	7	8	9	10
Difference from mean	·16	·03	·11	·25	·17	·04	·03	·36·	·07	·06

Mean of the differences = ·128 cm., i.e. nearly ·13 cm.

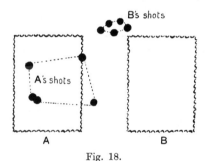

Fig. 18.
Shots at a postage stamp

It will probably be asked " Why should we find the difference from the mean and not the difference from the correct result ?"

The answer is that there are other causes at work besides those producing mere inaccuracy. The following illustration may help to make this point clear. *A* and *B* fire with a miniature rifle at a postage stamp stuck on a board, giving five shots each.   The results are given above.   (Fig. 18.)

B's shots are all close together but he has missed every time, and every time in the same direction.

A has three and a half hits and one and a half misses, but his shots scatter over a wider area than *B*'s. Probably *B* is as good a shot as *A* but something, wind, or faulty sights, has caused him to miss every time in the same direction. In order to compare the two results satisfactorily we should like to know (1) how far from the centre is a point in the middle of all the shots and (2) what is the average distance of each shot from this point. In the same way in using the method of average error we first find the mean of all the results. This may be compared to finding a point in the centre of all the shots. In the case of drawing lines it is 4·85 cm. and we conclude that some cause was acting all the time which tended to make all the results too small. Secondly we find the average amount by which the results differ from the mean. This corresponds to finding the average distance of the individual shots from a point in their centre, and just as this average distance would measure the accuracy of the shooting, so the average of all the differences gives a measure of the accuracy of drawing the lines.

For fuller information on this subject reference should be made to books on statistics mentioned in the appendix.

EXERCISE 23. *To test the accuracy with which distances moved by the elbow joint can be judged.*

The last experiment was arranged to test the muscles which move the thumb and forefinger ; in this one it is designed to test those which bend the arm at the elbow.

Take a post-card and cut four slits in it as shown in Fig. 19.

Cut out two strips of centimetre paper 1 cm. wide

and 30 cm. long and number each from below upwards beginning 10 cm. from the bottom. Pass these through the slits in the card as shown. The apparatus will work better if a second post-card is pasted on the back of the first by its edges, so as to keep the strips flat but at the same time allow them to slide easily. To use the

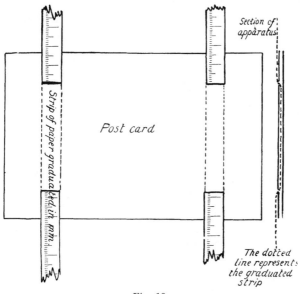

Fig. 19.

Apparatus for judging extent of movements made by an elbow joint

apparatus pin it to the wall and draw both slips of paper up till 0 is level with the upper slit in each case. Fold one of them over 10 cm. above the slit. Now close the eyes and draw this slip down till the fold catches on the card. Draw the other slip down through what feels to be the same distance. The number which is now opposite the top slit on the second slip shows the distance which was felt to be the same as that through which the first slip was moved. Have this result noted by someone else ; re-set at 0 and do the experiment ten times. Calculate the mean of the ten results and find the average difference from the mean as before. Do not examine your results till all ten are complete.

Pressure which can be exerted when a muscle contracts.

We have seen that besides obtaining information from movements as regards the distance moved we are also able to observe the resistance overcome. To lift 2 lbs. through a certain distance feels different from lifting one. Without going into the question of the exact source from which these sensations come we proceed at once to a few simple experiments.

EXERCISE 24. *To see what is the greatest force which can be exerted by a single set of muscles.*

Procure a hanging spring balance graduated up to 25 lbs. and fix it to the clamp of a retort-stand. Arrange the balance in such a position that its hook and the base of the stand can be comfortably gripped at the same time by the thumb and forefinger. Press the thumb and forefinger together as hard as possible and note the greatest pressure recorded by the balance. Exert a steady pressure, i.e. do not jerk. It is useful to lay the retort-stand on its side as this prevents the possibility of dragging the spring balance downward with the arm. Repeat the experiment ten times but allow intervals between the trials. Why ? Take the mean of the results. Do you find that any other muscles besides those you are using also contract at the same time ?

The instrument used in this experiment is the same in principle as that sold as a dynamometer. Fig. 20 shows a rough dynamometer made in this manner.

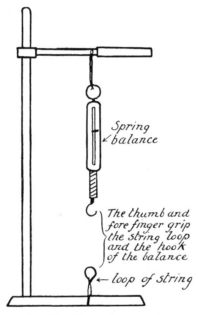

Fig. 20. Rough dynamometer

EXERCISE 25. *To see with what degree of accuracy a certain standard pressure can be repeated.*

For this experiment a somewhat more accurate form of balance than the last is necessary. It is best either to make a spring balance with a strip of elastic and graduate it in grams by actual experiment or use one of the so-called household balances in which the pressure on the spring is indicated on a circular dial in front. If the latter is used a board is required as a rest for the

arm.   The rest should be about 2 ft. long and a hole
for the finger should be made 10 ins. from one end.   It
can then be supported in a convenient position above
the balance by two retort-stands.   (See Fig. 21.)   To use
the apparatus lay the arm on the board and pass the
finger through the hole.   Press on the balance till the
indicator shows a pressure of 1 lb.   Then close the eyes
and repeat the pressure as exactly as possible.   Have
someone to make a note of the reading without telling

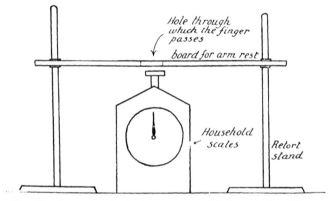

Fig. 21.   Apparatus for testing accuracy of judgment of resistance

you the result.   Repeat the experiment ten times.   Calculate the mean and average difference from the mean
as before.

Weighing without a balance.

Weights may also be compared by means of the
"muscular sense" or feeling of pull on the muscles and
resistance to movement, as when with eyes shut we hold
a 4 oz. weight in the left hand and a letter in the right
hand to find which is heavier, and if uncertain change

hands and try again before deciding whether the packet
will go by post for a penny.

An exacter test may be made with some known
weights hidden in pill-boxes. Cussons' weights of
$\frac{1}{10}$ lb. = ·1, $\frac{1}{20}$ lb. = ·05 and $\frac{1}{100}$ lb. = ·01 are very convenient for the purpose. The class works in pairs.
Each pair have two pill-boxes marked *A* and *B* respectively. One boy loads the boxes with ·10 in *A* and
·11 in *B* and hands them to the other who tries to tell
which is heavier; he writes down his judgment, then
opens the boxes to see what they really contain. Try
each test six times. The record may be kept like
this :—

Which box seemed to be heavier?		What the boxes really contained		Right or Wrong
A	*B*	*A*	*B*	
heavier	lighter	·10	·11	W
lighter	heavier	·10	·11	R
lighter	heavier	·11	·10	W

If your answers are generally right, next try whether
you can distinguish ·20 from ·20 + ·01. If the answers
are generally wrong, try whether ·05 can be distinguished
from ·05 + ·01.

It is very useful in a chemical laboratory to have
some idea of the weights of chemicals used without
testing them on the balance. Use the pill-boxes again.
In one of them put a 20 grm. weight. Put sand into the
other until the two boxes feel equally loaded. Then
use a balance and find what weight of sand has actually
been taken. The sand may be weighed whilst still in

the box, being partly counterpoised by an empty box in the opposite pan.   Record thus :—

>Intended weight of sand = 20 grams,
>Real weight of sand   = 21·52 grams.

If you now divide by the intended weight, thus

$$20\overline{)21{\cdot}52}$$
$$1{\cdot}076$$

the quotient will show the error ·076 as a decimal of the quantity estimated.   In the same way try to estimate 10 grams and 50 grams, and find the error after dividing by 10 and by 50 respectively.

Galton's weights.

The power of judging the energy with which a muscle contracts and the resistance offered to the movement has also been tested by an apparatus described by Sir Francis Galton in *Inquiries into Human Faculty* from which the following account is taken.   It depends upon the fact that in order that two weights may be felt as different the difference between the two must be a certain fraction of the smaller one.   This is exactly similar to what we have already seen to be the case with sensations of pressure on the skin, the apparent lengths of lines in contact with the skin, etc. and is an example of what has earlier in this book been called the Weber-Fechner Law (see p. 29).

The apparatus used by Galton, which is described in full in the appendix to the work quoted above, consists of a number of sets of three weights each, all exactly alike to the eye.   Suppose A, B and C are the weights in one set and A is the lightest, then it is arranged that the difference between A and B is a certain fraction of

A, and that between *B* and *C* is the same fraction of *B*. The same relation holds between the three members of each set, but the sets differ from one another in the size of the fraction. The method in which the weights are used is this. A set in which the weights differ by a large fraction is shuffled and handed to the person to be tested, who arranges them in what seems to him the order of weight and returns them to the person making the experiment. The experimenter by a system of secret marks finds out whether the weights have been correctly placed or not and hands to the person to be tested a second series in which the difference between the weights is smaller. Other sets with a still smaller difference are also used until a set is reached in which two mistakes in ten trials occur. The size of the fraction by which the weights differ in that set which are not correctly arranged gives a measure of the sensibility. If the fraction is large, the person is not so sensitive as he is if the fraction is small.

Fatigue.

EXERCISE 26.   *To illustrate the meaning of fatigue.*

Take a spring balance fitted up for measuring the greatest pressure which can be exerted by the muscles of the thumb and forefinger (Ex. 24, p. 56). Make a series of contractions in as rapid succession as possible, each as great as is possible. Note the results obtained. The latter results will be less than the first one. This is an illustration of a familiar fact that when a certain action has been performed a certain amount of energy has been used up and the same action cannot be repeated with the same intensity as before. The condition which is produced in this way is called **fatigue**.

Measure the decrease in the force of grip between the first and the last of the series of experiments made above.

In more careful studies of fatigue the indicating point of the spring balance is fixed to a writing instrument which makes a line on a sheet of smoked paper covering a revolving drum. Each contraction of the muscles then makes a mark on the paper, the highest point of which indicates the resistance overcome. Fig. 22 is a picture of a diagram made in this way. Examination

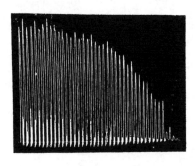

Fig. 22. Mosso's diagram

of it shows that when a series of contractions is made the tops of the curves gradually become lower and lower as fatigue comes on.

When a rest is allowed the fatigued parts very soon recover and the whole tracing can be repeated.

Fatigue of the senses.

What has been shown to be true of muscular action can also be shown to be true of many purely nerve processes as well.

EXERCISE 27. *To illustrate fatigue of sight.*

Stare fixedly with one eye at a large extent of patterned wall-paper or a large sheet of graph-paper held near the eye. Notice whether any part of the paper gradually becomes blurred.

Repeat the experiment in a faint light and see whether any part actually vanishes.

Edgar Allan Poe says " It is possible to make even Venus herself vanish from the firmament by a scrutiny too sustained, too concentrated, too direct[1]." Try the experiment with Venus or a bright fixed star and see whether you agree with him. The author has succeeded in the experiment with Jupiter at the end of an unusually fatiguing day's work.

EXERCISE 28. *To show fatigue of the sense of smell.*

Bring a bunch of pinks or other strongly scented flowers into a room and note the way the scent gradually becomes less noticeable as time goes on. Go out of the room for a few minutes and note that the scent acquires its original strength after one has been for a few moments in the open air.

EXERCISE 29. *To illustrate fatigue of the sense of touch.*

Determine many times in succession the smallest weight perceptible on some part of the hand. Is there any tendency for the smallest weight noticed to become gradually larger after the experiment has been repeated many times ?

Many similar experiments might be added showing fatigue of various senses, but enough have been described to serve the present purpose.

---

[1] *Murders in the Rue Morgue.*

The facts which have just been observed have a most important bearing on any experiments made on any of the senses. In testing various persons or various parts of the body, if the results are to be of the slightest value for purposes of comparison, it is absolutely necessary that the persons or parts tested should not be in a condition of fatigue. It is even unsafe to make a long series of experiments at once and then take the mean; if this is done the last results will be affected by fatigue and the mean will be unreliable.

EXERCISES.

1. Compare by any of the methods of this chapter the accuracy with which your right and left hands will work.

Are you right or left handed?

2. Repeat Exercise 24 and find out how the result is affected by (*a*) reading aloud while the experiment is going on, (*b*) holding the mouth wide open, (*c*) setting the teeth and clenching the other hand.

3. Find the time it takes you to run 100 yards (*a*) with and (*b*) without running corks. Do not run one 100 yards immediately after running the other, but do the two experiments at the same time on different days. Why?

4. Why does a man make faces when he jumps?

5. Invent a method for finding how accurately the foot can move at the ankle joint and at the knee.

6. Take a ruled line 1 in. long and run a pencil along it. Close the eyes and draw lines which seem to be 1, 2, 3, 4, 5, 6 ins. long respectively. Repeat several times and find the mean and average difference from the mean for lines of each length.

7. On a long railway journey in winter a passenger enters a carriage whose occupants have travelled in it for an hour and a half. He complains of a stuffy smell in the air which none of them notice and opens the window. They complain of cold which he does not notice. How is this?

8. Why is it difficult to carry a saucer of water full to the brim without spilling any ?

9. "If you eat too much sugar it begins to taste like salt." Why is this ?

10. In working through this chapter an experimenter tried Exercise 22 twice over, then Exercise 24, then Exercise 22 again. The average errors he got for Exercise 22 were as follows :

1st trial	...	...	...	·128 cm.
2nd trial	...	...	...	·204 cm.
3rd trial (after Ex. 24)		...		·562 cm.

How do you account for these results ? Try the same experiment on yourself.

11. Using a sharp pencil or fine pointed pen draw a number of fine lines nearly parallel and as close together as possible. Use a ruler to measure their average distance apart. Can you see lines only $\frac{1}{100}$ inch apart ? Can you draw lines $\frac{1}{100}$ inch apart ?

12. Sometime when you are on a ship or train try whether you can hold a cup nearly full of tea without spilling it. First try whilst you watch the cup. Then try without looking at the cup, but attending to the muscular sense, and gently giving way to it as the cup pulls your hand to right or left. If you succeed try to walk across the deck or carriage carrying a cup of tea.

CHAPTER VII

TASTE AND SMELL

Structure of the nose and tongue—Touch and taste distinguished— Connection between taste and smell—Just perceptible tastes

Apparatus and materials required.

Sugar, salt, clean handkerchief, powdered chalk, powdered marble, half a litre to a litre of water in

which a few cloves have been boiled. Solutions of sugar of various strengths. Flasks. Measuring flasks, 1000 c.c. and 100 c.c. Pipettes, 10 c.c. Glass tubing. Concave mirror.

The nose.

Before starting the work of this chapter it will be well to revise those parts of Chapter II which deal with the structure of the nose (p. 9).

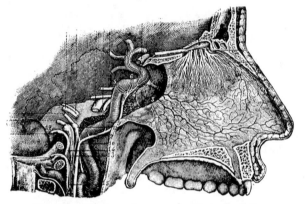

Fig. 23. Structure of nose and nerves of smell

It should be noted that the nasal chamber opens in front by the nostrils and behind into the gullet. During life it is divided into right and left halves by a plate of cartilage, and its two chambers are again subdivided by twisted scrolls of bone called **scroll bones**. These are covered with a thin skin and during life are kept continuously moist. Over the uppermost of the scroll bones branch the ends of a set of fine nerves, which find their way out from the front of the brain by way of the perforations described in Chapter II (p. 9). It is with the ends of these nerves that the actual process of smelling is carried out. The other scroll bones, the dividing partition of the nose and the inside lining of the nostrils are also supplied with nerves, but they are not nerves of smell and though a touch can be felt in these parts scents as such are without effect upon them. The pungent sensation produced by ammonia is said to be in part due to the effect of the gas on these nerves.

The tongue.

Carefully examine the surface of the tongue in a concave mirror and make a drawing of the parts which can be seen.

The surface of the tongue is not smooth but is covered over with very small pimple-like projections to which the name **papillae** has been given. These are of three different kinds.

First. The most numerous sort are small projections which cover almost the whole surface of the tongue and give it a sort of velvety appearance.

Second. Among these are others which appear as small circular red dots. They are scattered irregularly over the front of the tongue and are more cushion-like than the first kind referred to.

Third. The last kind is not easily seen except in a tongue which has been removed. They lie right at the back and are only just visible when the tongue is pushed out to its fullest extent. They can be made out with difficulty by pushing a bright teaspoon down the throat in such a way that the upper surface of the tongue is reflected in its bowl. The spoon should first of all be warmed with hot water to prevent irritation of the throat and to prevent its becoming steamed over with the breath.

5—2

These papillae are much larger than those seen
before—probably at least ·1″ across. Each is surrounded by a small circular trench and they are very
few in number.

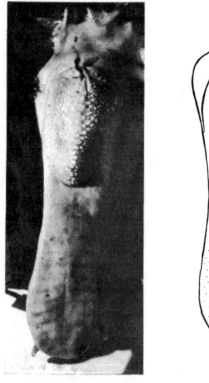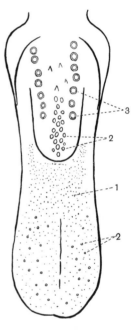

Fig. 24.   Sheep's tongue

All three kinds are easily seen on the tongue of a
sheep.   The only important differences between the
sheep's tongue and the human tongue are (1) that of

size and shape, (2) the fact that the sheep's tongue is entirely covered with a horny coating so as to prevent injury by the sharp edges of grass leaves. A figure of the sheep's tongue is given showing the three kinds of papillae numbered 1, 2 and 3.

Muscles of the tongue.

Experiment will show that there are two ways in which the tongue can be easily moved. (1) It can be moved as a whole, and (2) parts of it move in relation to one another. In response to the order " put out your tongue " the organ moves as a whole. The observer can then see that no part of the tongue is held stationary but the parts are in continual movement on one another. Try the experiment with a mirror and your own tongue. From the possibility of these two kinds of movement we should guess that there were two sets of muscles in the tongue, (1) connecting the tongue as a whole to the surrounding parts, (2) uniting various parts of the tongue to one another. Dissection shows both these different kinds of muscles to be present.

Distinction between touch and taste.

Place a small pinch of powdered marble upon the tip of the tongue. The sensation which follows is similar (except in place) to that which follows when the powder is placed on the hand. In other words the sensation is a sensation of touch alone.

Now in place of the powdered marble use a small pinch of sugar and compare the results. A sensation of touch, as before, is present plus something else which is expressed by saying that the sugar is sweet. This "something else" is what is called the sensation of taste.

EXERCISE 30. *See if sugar can be tasted on a dry tongue.*

Wipe the tongue dry with a clean handkerchief and see if pinches of salt, sugar, and powdered marble can be distinguished when placed upon it. What conclusion can you draw as to the connection between the solubility of substances and their taste?

Connection between taste and smell.

Most things swallowed are not only tasted but also smelt in the process. As a consequence a number of so-called tastes are almost or wholly due, in reality, to the sense of smell. A good example of such a "taste" is that of a weak infusion of cloves.

EXERCISE 31. *To illustrate the connection between certain tastes and smells.*

Boil one or two cloves in half a litre of water for a few minutes. Allow the infusion to cool. Then taste it. It appears to have a distinct and not unpleasant taste. Now take a clean glass rod and with it place a drop of the clove-water upon the tongue while the nose is held. Repeat this several times using sometimes clove-water and sometimes pure tap water. Carefully clean the rod after each test. Why? Keep a record of the numbers of right and of wrong answers.

Arrange the results in a table thus :

Substance used	Answer given	Right or Wrong or Uncertain

Percentage of " Right " answers =

Repeat the experiment and table again without holding the nose and again find the percentage of "Right" answers. Place a drop of clove-water on the tongue with the nose held and observe the way in which the taste suddenly appears when the nose is released and the vapour finds its way through.

Just perceptible tastes.

Just as we were able to measure the smallest weight which could be felt when placed upon the skin, so we can measure the smallest amount of any soluble substance which can be tasted.

EXERCISE 32. *To see what is the smallest amount of sugar which can be tasted when dissolved in* 1 c.c. *of water.*

Weigh out 10 grms. of sugar, dissolve in water and make up the solution to a litre. Every cubic centimetre of this solution contains ·01 grm. of sugar.

Now measure out 10 c.c. of this solution, put it into a 100 c.c. flask and make up to 100 c.c. This solution will contain ·001 grm. of sugar per c.c. If instead of taking 10 c.c. of the original solution we had taken 20, our resulting solution would have contained ·002 grm. per c.c.

By the method which is here suggested solutions can rapidly be made up of any strength desired.

Make a series of such solutions containing ·01, ·02, ·03, etc., grm. of sugar per cubic centimetre. Find out in which of these a sweet taste can just be detected. Suppose it is found that the solution containing ·04 grm. of sugar per c.c. just tastes sweet while the one containing ·03 grm. does not. To find out more precisely the exact amount of sugar which could just be tasted it would next be necessary to make up solutions which contained ·031, ·032, ·033, etc., grm. of sugar per c.c. and find out which of these tasted sweet.

The amount of solution used in each test must be the same. A convenient method of making sure of this is to have a glass dipping tube with a mark on one side, and in use to dip the tube into the liquid just up to the mark, then closing the top with the finger. A separate dipping tube for each strength of solution will save much trouble. The mouth should be rinsed with water after each test has been made.

EXERCISES.

1. Devise a method similar to the last for finding the smallest amount of salt which can be recognised.

2. Using a camel-hair brush and the solutions which have been prepared, test the different kinds of papillae on the tongue to see whether they are all equally sensitive to tastes and whether they are all sensitive to the same taste.

3. Are there any parts of the mouth, lips, palate, etc., which do not taste at all?

4. See which is more delicate, the silver-nitrate test for salt or the test by taste.

5. Nearly all poisons have a disagreeable taste and yet people sometimes drink poison without knowing it. How is this?

6. Names for tastes are:—hot, bitter, acid, acrid, sour, sweet, tart, pungent, soapy, astringent, salt, alkaline.

Taste, cautiously, each of the following substances and select the best name for the taste of each:—vinegar, alum, washing soda, capsicum, aloes, buttercup stem, saccharine, blackthorn berries, very dilute ammonia 1:100 of water, dilute ferric chloride solution, nitre.

7. Examine and draw what is seen in a slice of tinned tongue. See if you can discover, (*a*) muscles which fix the tongue to other parts, (*b*) muscles running from one part of the tongue to another.

8. Moisten your finger with milk and let a cat lick it. Describe and explain what you feel.

9. Make a list of substances whose tastes change when you have a cold in the head.

10. How is it that many of the names applied to tastes can also be applied to smells ?

11. Place the wires from an electric battery on your tongue and describe (*a*) what you feel, (*b*) what you taste.

CHAPTER VIII

SOUND AND HEARING

How sound is produced and reaches the ear—The machinery of hearing ; the ear-drum and internal ear—How sounds differ— The smallest sounds which can be heard ; watch test ; Politzer's Hörmesser ; whispering test—Pitch—The shrillest sounds which can be heard

Apparatus and materials required.

Tuning-forks, pith ball on fine thread, other sound-producing instruments, air-pump and alarum clock, coco-nut and rubber membrane or tissue paper, sand, living or recently killed frog, ear-plugs made of glass rod covered with rubber tubing, watch, Galton whistle.

Living frogs may be obtained from T. Bolton, Esq., 25, Balsall Heath Road, Birmingham.

The Galton whistle is sold by Messrs Reynolds and Branson, Commercial Street, Leeds, or may be obtained from the makers.

An excellent model of the human ear on a large scale is supplied by Mr C. Baker, 244, High Holborn, London, W.C. ; price 20s.

How sound is produced.

Take a tuning-fork and strike it. Touch the tip of one of the prongs with the finger or the lip. The prong can be felt to be vibrating, i.e. flying backwards and forwards at a great rate.

Touch the end of a sounding fork against a small ball of pith hanging from a fine thread. Notice that the vibrating fork flings the light pith ball a great distance from it.

Repeat the experiment using instead of the tuning-fork, a bell, a stretched wire, a drum-head, a long spiral wire such as is used in mantelpiece clocks.

How the sound from the vibrating object reaches the ear.

Place a clock or a watch on a duster folded up beneath the receiver of an air-pump. Gradually exhaust the air and note the way the sound grows fainter and dies away. Let the air enter and the sound quickly becomes louder. This shows that the air plays some part in carrying sound.

To show that vibrations can be communicated by the air from one object to another.

Take a coco-nut and saw it in half so as to obtain two bowl-shaped pieces of shell. Remove the contents of the nut and make a hole through the bottom about 1″ in diameter. Over the open end stretch a piece of rubber membrane such as is used by dentists. Failing this a tightly stretched piece of tissue paper will serve equally well, while instead of the nut a lard bucket or a glass diffusion jar may be used.

To use the apparatus sprinkle upon the membrane a few grains of sand or even some small pith balls.

Now make a moderately loud sound near the opening in the bottom, e.g. get someone to sing into it. As a result the air in the bowl is thrown into vibration and this in turn moves the membrane at the top of the bowl. Its movements are only slight but they are made evident by the dancing of the sand grains or pith balls on its surface.

The receiver of a phonograph works on exactly the same principle. The air is thrown into vibration by the voice or some musical instrument and there is a glass, or mica, or metal disc which in its turn is thrown into vibration by the air. This disc carries a needle, which in its turn rests against a revolving wax cylinder on which the vibrations are recorded.

The machinery of hearing—The ear-drum.

The fact that we are able to hear depends upon the fact which has just been illustrated. In the human ear there is a stretched drum similar to, but smaller than, the one just referred to.

To examine the ear-drum of a frog.

In the frog and the toad the ear-drum is more easily seen than in other animals since it lies upon the same level as the surface of the body. Examine a living or recently killed frog ; look at the sides of the head just behind the eyes and notice two oval discs of smooth skin bordered with a slightly raised rim. These are the ear-drums. Using a dead frog remove the drum by cutting carefully round the edge of it with fine scissors. A passage can now be seen passing down into the head. Pass a fine bristle or a hair down this. Then open the frog's mouth and see where the bristle appears. Also notice a thin curved rod of bone which runs across from

the ear-drum to the outer surface of the skull. This
bone is named the **columella** and it serves to transfer
the vibrations of the ear-drum to the organ of hearing.

The internal ear or organ of hearing.

An animal cannot hear with its ear-drum alone.
There are persons whose ear-drums are perfect who are
deaf and others whose ear-drums are broken can still be
made to hear. The ear-drum and columella are simply
the means by which the vibrations of the air are conveyed to the internal ear.

EXERCISE 33. *To show that sounds can reach the
internal ear and be heard by other means than the ear-
drum.*

Close the ears with the fingers or plug with cotton-
wool and get someone to tap with a pencil on the seat
of a wooden chair or the top of a table. Hardly any
sound reaches the ear. Repeat the experiment, but
before tapping grip the chair back or edge of the table
tightly with the teeth. The sound can now be heard
easily. By what path do you suppose the vibrations
have travelled in this case? An instrument has been
made on this principle for enabling persons to hear, if
their ear-drums are broken, or if the small bones which
transmit the vibrations to the nerve of hearing fail to
act properly. The instrument consists of a sheet of
ebonite in the form of a fan which catches the vibrations
of the air. The edge is then gripped with the teeth and
the vibrations reach the internal ear by the same means
as in the last experiment.

The dissection of the internal ear of any mammal is
a very difficult matter except for those who have had
a good deal of experience of such work, as the parts lie

embedded in a dense mass of bone, which has already
been referred to in the description of the skull (p. 10),
and they are extremely soft and fragile and consequently
easily damaged when the bone is removed. A useful
general idea of the relations of the parts may be

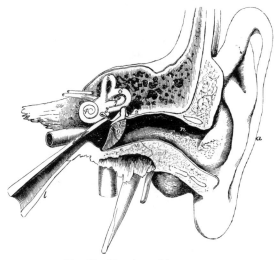

Fig. 25.   Structure of human ear

1 ⎫	*Chain of bones*	4	*Cochlea*	a	*Ear-flap*
2 ⎬	*going from*	5 ⎫	*Semi-circular*	n	*Ear passage*
3 ⎭	*ear-drum to*	6 ⎬	*canals*	l	*Eustachian tube*
	internal ear	7 ⎭		o	*Ear-drum*

gathered from the study of their relations in the dog-
fish or skate, where dissection is comparatively easy,
since the skull is cartilage and not bone, but detailed
instructions for carrying out such a dissection are
beyond the scope of this book[1].

---

[1] See *Practical Zoology*, Marshall and Hurst, p. 279.   Dog-fish
(*Scyllium canicula*) may be obtained from the Marine Biological
Laboratory, Citadel Hill, Plymouth.

The internal ear consists of cavities in the ear bone. Into these cavities the nerve of hearing passes through a small hole entering the bone from the inner side. On the outer side of the bone is a thin spot against which rests the end of a chain of bones which convey the vibrations from the ear-drum, like the columella of the frog.

The cavities are thus completely shut in. In form they consist of a set of three tubes (see Fig. 25), **semi-circular** in shape, one horizontal and the other two vertical in position. Communicating with these is a fourth tube coiled into a spiral like a flattened snail shell. The whole apparatus is filled with fluid and into it project at certain points the hair-like endings of the nerve of hearing. The actual process of hearing apparently takes place in the spirally coiled tube which is named the **cochlea** (which means snail shell). The semi-circular tubes seem to act as a kind of delicate spirit level and help to keep us informed of the position and movements of the head. If, in any animal, they become damaged or diseased the animal is no longer able to keep itself in an upright position, but perpetually tumbles down and stumbles up again until exhausted.

The ear-flap.

So far nothing has been said of the ear-flap. This serves the purpose of gathering the air-vibrations together and directing them upon the ear-drum.

EXERCISE 34. *To show how the ear-flap affects hearing.*

Place a watch in such a position that its ticking is only faintly audible. Place one hand behind the ear and see whether the sound becomes any louder. Repeat the experiment, but instead of the hand use a large sheet of paper folded into a cone with the pointed end in the ear.

Do the results of this experiment suggest any explanation of (1) the use of a speaking-tube as a substitute for a telephone, or (2) the use of an ear-trumpet to the partly deaf? Would an ear-trumpet be of any use to a person whose ear-drums were broken?

Information supplied by the sense of hearing.

How do sounds differ from one another? Think of some of the sounds one hears frequently. What difference is there between the ticking of a watch and hammering on an anvil? between an air sung by a bass and the same air sung by a soprano? between an air played on the piano and the same notes played on the violin?

The answers to these questions will suggest some of the differences in sounds. The following experiments will make the differences still clearer.

EXERCISE 35. *To illustrate three ways in which sounds differ from one another.*

Sound a tuning-fork and hold it at arms length. Bring it quickly close to the ear. The sound becomes **louder.**

Sound an *A* fork and a *C* fork. Neglecting accidental differences in loudness, a difference is also noticeable in that one fork has a shriller sound than the other, i.e. it has a different **pitch.** It is precisely this difference which is indicated by the letters *A* and *C* applied to the forks.

Finally compare the note given by a tuning-fork with the same note struck on a piano. The notes have the same pitch and yet there is still a difference, neglecting any differences in loudness. This difference is not very easily put into words. It is generally explained by saying that the two notes have each a certain quality called **timbre** of their own. The name **clang-tint** is also sometimes used.

In making any experiment with sounds a great difficulty is met with in avoiding distracting noises. For this reason the experiments here given are extremely rough in character and serve to suggest methods rather than to give exact results.

The loudness of sounds.

EXERCISE 36. *To see how far a watch must be removed from the head until it ceases to be heard.*

Remove a watch slowly from the side of the head of the person tested keeping it in a straight line running directly outwards from the ear. Note that the ticking gradually becomes fainter and finally ceases to be heard. The ear which is not being tested should be closed with the finger or a plug (convenient plugs can be made of glass rod over which rubber tubes are slipped). Repeat the experiment with the other ear. Record the distances at which the watch ceases to be audible in a set of five trials with each ear. Take care that the ear does not get tired by repeating the experiments in too rapid succession. Do you find both ears equally sensitive?

The smallest sounds which can be heard.

The power of hearing very small sounds can sometimes be tested by various accidentally produced noises. If the room is still and a fly settles on and runs across

the book you are reading, can you hear a very faint
rustling ? Can you hear the very soft thud that is
produced by the fall of a small pinch of ashes (not
cinders) from a dying fire ? What is the faintest sound
which ordinarily catches your attention ? Can you hear
a Bunsen burner when burning "silently"? Can you
hear a Bunsen when the gas is turned on but not lit ?

The use of the watch as described in the last
experiment for testing the faintest audible sound is
open to the objection that persons tested with different
watches cannot be easily compared. If, however, a
number of persons are tested with the same watch and

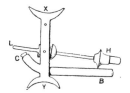

Fig. 26.   Politzer's Hörmesser
X, Y, *Holder.*   H, *Weight.*   L, *Lever.*   B, *Block*

among them one or two require to have it *much* closer
to their ears than the rest, it is safe to assume that they
are at any rate slightly deaf and they should consult
a medical man on the subject.

Tests which give an absolutely uniform result have
been devised, e.g. it has been suggested to make a
standard sound by dropping a ball of cork of known
weight from a given distance on to a glass plate and to
find at what distance from the ear it ceases to be heard.
An instrument for producing a sound of a certain
standard loudness can be obtained from J. T. Raper,
50, Great Russell Street, London, W.C. It is named
Politzer's Hörmesser. (Fig. 26.)

It consists of a weight which is carried on the end of a lever. It can be raised to a certain height above a block upon which it is then allowed to fall. It is recommended for the tests suggested by the British Association Anthropometric Sub-committee.

EXERCISE 37. *To test the acuteness of hearing by whispering.*

Find a distance at which a *normal* ear can just hear a whisper. Place yourself at this distance from your teacher and write down as many of the numbers dictated by him in a whisper as you can hear. Any five of the ten groups of ten may be selected. Find out how many of the 50 numbers you were unable to hear or wrote down incorrectly.

Andrews' Table of numbers for testing acuteness of hearing by whispering.

I	II	III	IV	V	VI	VII	VIII	IX	X
6	84	19	90	25	14	8	52	73	24
29	69	53	7	13	31	93	35	41	95
42	17	34	39	46	9	27	64	16	62
87	92	28	62	7	65	60	81	95	49
53	33	97	84	54	98	15	6	57	80
94	26	45	21	70	76	74	19	38	71
70	50	72	56	91	40	36	78	20	16
35	75	60	75	83	23	49	40	89	3
18	48	3	43	68	52	82	23	64	58
61	1	86	18	92	87	51	97	2	37

As in the case of the watch this cannot be taken as an absolute test, because it is very difficult to be sure that the words are always whispered under exactly the same conditions, but out of a number of persons tested under the same conditions those with weak hearing can be picked out.

Note for Teacher. These groups of numbers are selected according to the result of a careful series of experiments on the audibility of numbers. Those interested in the subject should refer to the original papers by B. R. Andrews : "Auditory Tests," *American Journal of Psychology*, xv (1904), pp. 14—56 and xvi (1906), pp. 302—326.

It might be useful to dictate these figures in a whispered voice to the whole class, dictating from different parts of the room so as to give everyone an even chance of hearing.

Pitch.

It is found that the difference in sounds which is called **pitch** really depends upon the rapidity of the vibrations of the object producing the sound. Slow vibrations give a low note, rapid vibrations a high one. If the vibrations are slow enough they are not heard as a musical note at all, but as a patter of separate sounds. The number of vibrations which must occur in a second to produce a musical note varies with different people. It has been measured by using large and slowly beating tuning-forks.

As in the case of the loudness of sounds, the power of distinguishing between notes of different pitch can be roughly tested by means of sounds which occur in one's ordinary experience. Can you e.g. distinguish between a note on the piano and the ones half a tone above and below it? Can you tighten a violin string till it has the same pitch as the corresponding note on the piano? Do you notice when a piano gets out of tune, if the notes are struck one at a time?

Shrill notes.

Very shrill notes cannot be heard at all. A bat's squeak cannot be heard by many people and may be used as a rough test of the highest note that can be heard. Others cannot hear the buzz of a mosquito or the chirp of a grasshopper.

The Galton whistle.

An instrument has been designed by Sir Francis Galton, whose name has already been mentioned in connection with weight tests (p. 60), for testing the power of hearing very shrill notes. It has been described in *Inquiries into Human Faculty* and elsewhere. It consists of a whistle of fine bore with a plunger in one end which shortens the length of the tube. A puff of air is sent through it by means of a rubber bulb and tube. In use a note is first sounded which can be easily heard. The pitch is then raised by turning a screw till a note is reached which is not heard. The instrument is provided with an indicator showing the number of vibrations per second for each position of the screw.

EXERCISE 38.   *To find out what is the shrillest note you can hear.*

If a Galton whistle[1] can be obtained, have it sounded in the presence of the people to be tested and get each of them to make a note of the point at which they cease to hear it. Also have recorded at the same time whether they can hear a bat's squeak and a mosquito.

---

[1] Galton whistles are supplied by Messrs Reynolds and Branson, Commercial Street, Leeds.

EXERCISES.

1. A glass marble dropped on a slate bounces more and more rapidly. At first you hear a series of taps, but later, just before it stops, it makes a kind of squeak. How is this ?

2. Strike a tuning-fork, close your ears with a pair of plugs, and place the handle of the fork on the top of your head. Notice the way the sound appears. How do you suppose it reaches the internal ear ?

3. Close the ears as before and take a pencil between your teeth. Press the end of the pencil on a watch and see whether you can hear the ticking.

4. Spread a piece of thick cloth on the bench and clamp a metre rule vertically beside it. Test your acuteness of hearing by sitting 5 feet away and getting someone to drop a pin on to the cloth from various gradually increasing heights until you just hear the faint thud that the pin makes on striking the cloth.

5. Hold a watch exactly one metre from the ear. Then suddenly move it 2 cm. further away. Notice if the ticking seems fainter. Replace it and take it suddenly 4 cm. further away. Repeat until a distance is reached which just makes the ticking fainter. In this way compare the power of distinguishing loudness of sounds possessed by various people.

6. Make for yourself from a sycamore twig a whistle of variable note. The bark is loosened by hammering it. The wood inside becomes the moving plunger.

CHAPTER IX

LIGHT

Rays and beams—Reflection ; objects which shine by their own and by reflected light—Bending of light by transparent objects—Lenses and their various forms—The pictures thrown by lenses

Materials and apparatus required.

Small lamp or candle, cards with hole in centre, mirror, sheet of note paper, wax vesta, gas mantle,

shilling and other common objects for Ex. 42, rect-
angular block of glass, pins, triangular block of glass
(prism), lenses or half lenses—set of six.

How light travels.

In beginning work on sound it was necessary to begin
by studying the way in which sounds are produced and
travel from place to place. So here it is necessary to
start by studying the way in which light travels.

EXERCISE 39. *To show that light travels in straight*
lines.

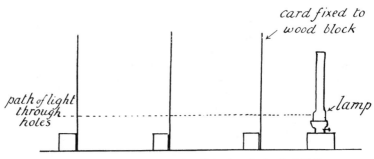

Fig. 27. Apparatus to show that light travels in straight lines

Take a lighted candle or a lamp and place in front
of it a card screen in which a small hole has been pierced.
Find out in what positions with regard to each other the
hole, the lamp and the eye must be placed in order that
the lamp shall be visible through the hole. (See Fig. 27.)

Take a second piece of card perforated in the same
way and arrange the lamp and the two cards so that
the light passes through both holes.

Make a diagram to show the way in which the light
has travelled from the lamp to the eye. This is not the

place for a full explanation of the theory of light but it will save words to give two short definitions.

1.   A very fine line of light is called a **ray**.

2.   A bundle of rays of light is called a **beam**.

Bending of light by reflection.

EXERCISE 40.   *To show that rays of light can be bent by reflection.*

Place a mirror in the position occupied by the eye in the last experiment, and see whether a reflection of the candle can be seen from any position in which the eye is not in a straight line with the candle and holes. In other words, can the beam of light from the candle be made to turn a corner?

Illustrate the result of the exercise by a careful diagram.

EXERCISE 41.   *To show that other objects besides mirrors can reflect light.*

Replace the mirror of the last experiment by a piece of white paper, and note the white spot formed by the light from the lamp.   Find out whether the white spot looks brighter from any position than from any other.

Most objects which are visible in the natural course of events are seen only by means of the light which they reflect.   Thus in reading this page, what the reader really sees depends upon the fact that white paper reflects light whereas printer's ink does not, or only does so to a very slight extent.   If the book is taken into a dark room both page and print are alike invisible, since there is no light for them to reflect.

EXERCISE 42. *To see whether various objects shine by their own or by reflected light.*

The fact that objects which shine by reflected light are invisible in the dark affords a ready means of distinguishing those which shine by their own light and those which do not. Find out which of the following objects shine by their own and which by reflected light : a lighted candle, a shilling, phosphorus, the moistened head of a wax vesta, an incandescent gas mantle with the gas not lit, the same with the gas lit, a piece of platinum wire or foil in the flame of a Bunsen, the roof of a greenhouse in the sunshine, lightning, sparks from a horses hoof on a flint road, a glow worm, the moon.

It has been suggested above that light always travels in straight lines ; this is not strictly correct as has already been seen in the experiments on reflection. The following experiments will show other ways in which bending may take place.

Bending of light in other ways.

EXERCISE 43. *To trace the path of a beam of light through a block of glass.*

Place a block of glass on a sheet of white paper and stick a pin into the paper about 2 inches away from the glass. Look at the pin through the glass from various positions. Does the stem of the pin as seen through the block appear in a different place from its head seen over the top of the block? Make a sketch of what can be seen from each position.

Next take a number of pins and stick them in the paper between the first pin and the glass, arranging each in such a position that it hides the last when viewed through the glass. In the same way put in some more between the glass and the eye on the near side. Since each pin hides the one behind it, each must be in the path the beam of light takes in coming from the first pin to the eye. If now the glass is removed and the pin pricks joined by pencil lines, we obtain a map of the path taken by the light from pin to eye.

How many times has the beam of light from the pin been bent in passing through the glass?

Is there any arrangement of pin, glass, and eye which allows the light to pass through without bending at all?

Repeat the experiment using a "prism," i.e. a flat-sided block of glass with triangular ends.

How many times is the light bent in this case and in which direction?

Light bent in this way is said to be **refracted**.

Lenses.

EXERCISE 44. *To examine and describe some common lenses.*

The name **lens** is given to a curved and polished piece of glass. The name is derived from the Latin *lens,* which means a lentil, the reason being that the commonest sorts of lenses are much the same shape as lentils.

Lenses are of two sorts; **convex** lenses, which are thicker in the middle than at the edges, and **concave** lenses thicker at the edges than in the centre. Fig. 28 shows sections through six such lenses.

Examine, describe and draw imaginary sections through as many lenses as can be obtained, e.g. magnifying glasses, spectacle glasses, bull's-eye lantern, etc. How many different kinds of lens are possible?

Exercise 45. *To show the way in which a lens bends the rays of light which pass through it.*

Use a common convex lens. Place the lens between a lighted candle and a sheet of white paper. Place the paper at first close to the lens and then gradually draw it further and further away. Note the changes that occur as the paper is drawn further from the lens. At first there is a circle of light nearly as large as the lens.

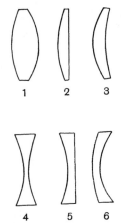

Fig. 28. Sections of lenses

1, 2, 3. *Convex lenses*
4, 5, 6. *Concave lenses*

This gradually gets smaller till a picture of the candle flame appears on the paper. If the paper is pushed still further back the picture fades and the light gradually becomes fainter and also forms an enlarged patch upon the paper. The point at which there is a sharp picture on the paper is called the **focus** of the lens.

Can anything be seen in the picture besides the flame of the candle? Bring various objects near to

Light   91

the flame and see whether they also appear in the
picture, e.g. try a sheet of wire gauze held just in front
of the flame. Also try the effect of covering in the
distance between the lens and the screen with an arch
of brown paper.

Place close in front of the lens a piece of brown
paper with a hole cut in the centre. What effect does
this produce as regards size, brightness, and distinctness
of the picture? For full details of the way in which a
lens makes a picture by bending the rays of light, reference should be made to a text-book of Physics (e.g.
Watson's *Elementary Practical Physics,* pp. 147—154).
The purpose of these experiments will have been served
if they succeed in bringing out the following facts :

1. A lens throws a picture upon any suitable screen
of all the objects in front of it, whether they shine by
their own or by reflected light.

2. In one position of the screen the picture is sharp
and distinct ; if the screen is nearer to, or further from,
the lens than this the picture is blurred and indistinct.

3. If the size of the lens is reduced by covering all
but a small central part with an opaque substance such
as brown paper, the picture is reduced in brightness but
not in size. At the same time the sharpness of outline
is increased.

EXERCISES.

1. Repeat the above experiments with lenses of as many different
kinds as possible.

2. Place a penny on the bottom of a photographic developing
dish and hold it in such a position that it is just concealed by the
edge of the dish. Fill the dish with water. Describe and explain
what changes you see.

3. See if a picture can be obtained by using a round-bottomed flask filled with water in place of the lens of the last experiment.

4. Carefully examine a camera which has a ground-glass screen at the back. Compare with the apparatus used in Exercise 45.

5. Hand cameras often have a small piece of ground glass let into the top in which the view which is being photographed can be seen. Examine one of these "view-finders" and compare with the apparatus of Exercise 45.

6. What do you see when you try to look through a dewdrop towards the landscape beyond?

7. If a magic lantern is available the properties of different lenses may be tested by holding them in the beam of light. The track of this beam may be made visible by filling the air with smoke, e.g. from two adjacent basins one containing strong ammonia solution, the other salt and sulphuric acid. It is useful to have some cards of the size of lantern slides and cut with apertures of different shapes and sizes, e.g. slits, circles and crosses. For details of such experiments, see *Optical Projection* by Lewis Wright, published by Longmans, Green and Co.

CHAPTER X

THE EYE

The eye compared with a camera—The external features of a sheep's eye—The internal parts of a sheep's eye—The picture thrown by the lens—How to see the back half of a living eye

Materials and apparatus required.

Sheep's or bullock's eyes from the butcher's. Dissecting instruments (one pair of scissors, one pair forceps and one scalpel for each two pupils). Pill boxes and cardboard tubes from incandescent mantles. Tissue paper. Coverslips. Short piece of penholder.

Enlarged model eye from Mr C. Baker, High Holborn, London, W.C., price 15s., or larger model, price £1.

The structure of the eye.

In the eye are a lens and a screen just as in the apparatus made in the last experiment. The eye is in fact a small camera. We now proceed to examine an actual eye and see if possible how it works. The most convenient eyes for examination are those of the sheep or ox, which can usually be obtained in quantity from any town butcher.

Dissection of the eye of the sheep. (1) Externals.

What is written here of the sheep's eye is also true of the eye of the ox and will also apply, with purely trifling differences, to the eye of the rabbit.

Place the eye to be examined on the desk and turn it into the position it would occupy in nature. Note its roughly globular shape. It is convenient to use the same names in describing the parts of the eyes which are used in describing the terrestrial globe.

Thus we can distinguish an **axis** with two **poles**, one forward called the **anterior** pole the other backward the **posterior** pole. The **equator** would divide the eye into front and back **hemispheres**. The words **inner** and **outer** are on this account used to mean towards the centre or towards the circumference of the eye-ball. Thus, if the eye is looked at from behind, the surface seen would be called the *outer* surface of the posterior hemisphere. The front half of the eye is covered with an arched transparent covering called the **cornea**. In front of this, and so closely attached to it that it can only be separated by dissection, is a thin transparent skin. This is really continuous at its edges with the skin of the inside of the eyelids and is named the **conjunctiva**.

In passing toward the equator of the eye the horny cornea gradually gives place to a tough fibrous coat commonly spoken of as the "white" of the eye. To this the technical name of **sclerotic** is given and to it are fixed the muscles which serve to roll the eye about into various positions, so that various objects may be brought into view.

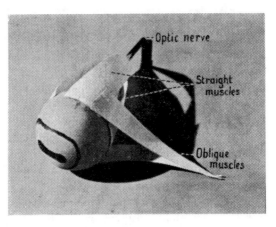

Fig. 29

External view of a model eye-ball made of a tennis ball, paper and a penholder

The whole of the back of the eye as it appears when recently cut out of its socket is completely covered with a mass of muscle and fat, while somewhere not far from the posterior pole is seen a white cord which is the **optic nerve** entering the eye-ball.

The muscles just referred to can be dissected out with care. The fat should gradually be removed from them and their points of attachment to the sclerotic made out. If this is done they will be found to be six in number. Four of them, called **straight** muscles, are attached round the equator, which they divide into four roughly equal parts. They lie, roughly speaking, above and below the eye-ball and in two positions at right angles to this. At their other ends they are fixed to the bony socket of the eye near the point where the optic nerve enters.

The remaining two muscles are named **oblique**. They are fixed to the eye-ball in the region of its equator, one above and the other below it, and their other ends are fixed to the bony eye-socket on the side toward the nose. (See Fig. 29.)

Dissection of sheep's eye continued. (2) Internal parts.

To make out the internal structure of the eye it is well to have two eyes and divide one into front and back and the other into upper and lower halves. The same result can be attained by allowing half the class to do one dissection and the other half to do the other. Failing either of these methods, it is best to cut a single eye first into front and rear portions and then divide the front half again in a direction at right angles to the first division. (See Fig. 30.)

Place the eye to be operated on in a dish of water and with a pair of sharp dissecting scissors cut the sclerotic right round the equator so that the eye is divided into front and rear halves. When this has been done an irregular mass of clear jelly will escape. This is called the **vitreous humour.** It is like rather tough jelly and keeps its general shape but is so extraordinarily transparent that it is only possible to distinguish it from the water in which it lies with the greatest difficulty.

Carefully examine the back half of the eye-ball. Notice that the whole of its inner surface is covered with a gray coating—the **retina**. This has a slightly fibrous appearance and the fibres run outwards from a spot at the back near its centre. These fibres are really the finely divided ends of the optic nerve, which

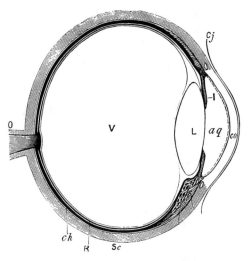

Fig. 30.   Structure of interior of eye

O, *Optic nerve*	V, *Vitreous humour*	Cj, *Conjunctiva*
ch, *Choroid*	co, *Cornea*	I, *Iris*
R, *Retina*	Sc, *Sclerotic*	aq, *Aqueous humour*
L, *Lens*		

has already been seen outside the eye-ball, and it is by means of them that the action of seeing takes place. Starting from the point of entry of the optic nerve, and running rather irregularly over the inner surface of the retina, are a number of fine branching **blood vessels.**

In the centre of the hinder hemisphere of the eye, i.e. at the posterior pole, is a spot called the **yellow spot**, so named from the fact that it is tinged a yellowish colour.

Outside the retina, i.e. between it and the sclerotic, is a layer of dead-black colouring material. This is called the **choroid coat** and extends over almost the whole of the inner surface of the eye.

Examine next the front half of the eye-ball. Traces can be seen of the retina which reaches forwards beyond the equator and ends in a finely notched edge.

In the centre of the front of the eye, i.e. toward the anterior pole, lies an almost globular glassy body—the **lens**. This is transparent but not so water-clear as the vitreous humour which fills the centre of the eye-ball. It will be noticed that the lens is enclosed in a fibrous bag which closely fits it and by means of which it is kept in place. This bag or **capsule** is fixed to a ring of muscular projections from the choroid coat at the front of the eye and by their contraction the shape of the lens can be altered. What change in the shape of the lens will be produced by contraction of this ring of muscles? The lens can now be removed by snipping carefully round the margin of the fibrous capsule with scissors. Remove it and place it on a glass slip and hold it over a page of print so as to test its magnifying power.

It is not very easy to show that it can throw a picture. A picture could be seen from outside on the back of a dead eye, but the black choroid coat intervenes and absolutely prevents any light from passing through. Now in albino animals the choroid is very faintly or not at all pigmented. It is said that the eye of an albino rabbit shows a picture from outside on the back or posterior hemisphere of the eye[1].

A simpler method is the following. Take a pill box large enough to slip over the end of one of the card tubes in which incandescent mantles are packed. Cut a circular hole in the bottom of the pill box and rest

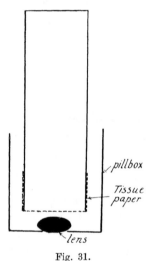

Fig. 31.

Apparatus to show that the lens
of the eye will cast a picture

the lens carefully in the hole. Over the end of the card tube fix a sheet of tissue paper or tracing paper. Hold the lens over a page of print in a well-lighted situation. Lower the card tube and tissue paper down close to the lens as shown in Fig. 31.

When the paper comes close to the lens a picture of

[1] The experiment is described by Sanford, *Experimental Psychology*, p. 89.

the printed page can be seen, usually somewhat distorted owing to the fact that the lens very easily loses its shape when removed from the eye.

The parts of the eye in front of the lens can now be seen. Most important of these and close to the lens is a curtain named the **iris**. At the back it is black in colour, but when seen from the front shows various colours in different eyes. The differences in colour of various eyes are in fact due to differences in the iris. In the sheep's eye it is commonly brown. In the centre of it is a hole called the **pupil**, which in human beings is circular, but in the sheep is a horizontal slit slightly widened at each end. Running round this hole in the margin of the curtain is a ring-shaped muscle by the contraction of which the hole in the iris can be partly closed. The mode of opening and closing of this aperture may be roughly compared to the closing of a bag by means of a running string in the neck of it.

The iris diaphragm of a camera is named from the iris of the eye because it has the same effect although it works in quite a different manner. The iris of the eye takes its name from the Greek Iris, the rainbow, the goddess, the many coloured flower.

The lens and fibrous capsule form the hinder wall of a cavity whose front wall has already been seen from outside. This front wall is the **cornea**. The cavity in life is filled with a clear watery fluid called the **aqueous humour**.

When all the parts described have been seen the following sketches should be made. They should be of a good size, not less than three inches in diameter :—

1. Drawing of the front half of the eye seen from inside.

2. Drawing of the back half of the eye seen from inside.

3. Section of the eye passing from front to back.

EXERCISE 46. *To examine the back half of a living eye.*

It is not possible under ordinary conditions to see in through the pupil of a living eye since the black choroid coat does not reflect any of the light which falls upon it. Some animals, e.g. cats, have a shiny iridescent layer

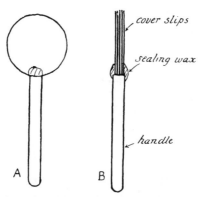

Fig. 32. Rough ophthalmoscope
(A, *seen from front;* B, *from side*)

(i.e. like shot silk) as well as the choroid, and this reflects the light well. Hence such animals' eyes seem to shine in the dark.

If a strong beam of light can be thrown straight into the eye, some light is reflected from the retina and this can be seen. An instrument named the **ophthalmoscope** is used for this purpose. The following is a description of a rough ophthalmoscope.

Procure four or five thin glass coverslips 1 in. diameter such as are used for microscopic work. Very carefully clean and dry them and fix them in a small wooden handle. Two inches from the end of a penholder will do very well. The end should be notched and the glasses fixed with sealing wax or seccotine. (See Fig. 32.)

The person whose eye is to be examined now seats himself beside and a little in front of a good oil lamp. The person using the instrument sits facing him as close as is convenient. The observer holds the glass close to

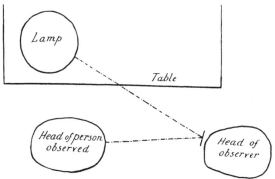

Fig. 33. The rough ophthalmoscope in use
(*The dotted line shows the path of light from lamp to the eye observed*)

his own eye in such a position that a beam of light is reflected into the eye to be examined and so that at the same time he can look through the glass. As soon as the correct position has been obtained the pupil will change from black to dull red. The observer now gradually approaches the eye and when a distance of rather less than four inches is reached the back of the eye will become clearly visible. At this point the back of the eye is in reality seen through its own lens.

In the centre of the dull red pupil will be seen a dark dot which is the point of entrance of the optic nerve. Radiating from this is a network of blood vessels not very easily seen. Make a drawing of what can be seen in this way. Fig. 33 above shows the arrangement of the lamp, ophthalmoscope, and person examined, which is necessary.

EXERCISES.

1. Why does the pupil of the human eye seem black ?

2. In white-haired animals (albinos) the eyes contain no colouring matter at all and yet the eyes are not white. How is this ?

3. Do cats' eyes shine in *absolute* darkness ; observe and explain.

4. Make a drawing of all that can be seen of your own eye with a mirror.

5. Stand beside a lamp and get a companion to look towards you from a distance of 12 ft. or more. Describe the appearance of his eyes.

CHAPTER XI

HOW THE EYE IS USED IN SEEING

Effects of pressure on the retina, and Le Cat's experiment—Accommodation—Changes in size of the pupil—Near point of vision—Defects in the eye—Long sight—Short sight—Tests for these and the principles on which they depend—Astigmatism—A note on the care of the eyes

Materials and apparatus required.

Cards, lamp or candle, pin, small mirror, 10 in. glass tubing, two corks, pins, large lens, candle and screen as in Ex. 45, optician's test cards, round bottomed glass flask and oval Florence glass flask.

The eye a camera.

If the structure of the eyes is compared with that of a photographic camera, it will be noticed at once that the two things are very similar. The lens casts an image upon the back of the eye, and the sensitive nerve endings of the retina take the place of the sensitive plate.

Now it has already been seen that the picture thrown by a lens is upside down. The same thing must obviously happen in the eye, and it has often been remarked as a curious fact that although this is so we still see things the right way up. The really remarkable fact is, that we are not commonly aware of a sensation as being "in the eye" at all, but invariably think of the object seen. So constantly is this the case that, if by purely mechanical means the nerves of one side of the retina are made to convey a message to the brain, it seems as if some bright object had been seen outside the eye and on the opposite side to the one stimulated. It is said that those who are born blind and have recovered their sight as adults have had to learn that what they see is not in actual contact with their eyes.

EXERCISE 47. *To show the effect of pressure on the retina.*

Close the eye and turn it in toward the nose. Press gently with the finger tip on the outside of the eye-ball; note the bright spot and ring which appear to be seen to the inner side of the field of view. Carefully describe the appearance seen.

EXERCISE 48. *"Le Cat's" experiment.*

A similar effect can also be produced in another way. It is possible to make light fall upon the retina in such a manner that the picture on it is the same way round and the same way up as the actual object. To illustrate how this can be done it is useful to look again at the apparatus used for showing how pictures are made by lenses (p. 90). Bring the lens of this close to a lamp and hold a white paper behind it, but too close

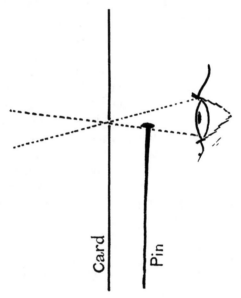

Fig. 34. Le Cat's experiment
(*The dotted lines show the course of light-rays
to the eye*)

for a picture to be formed. Now hold a pencil close in front of the lens. A shadow of the pencil appears on the paper screen, the same way round as the actual object. The reason for this is that the screen is too close to the lens to allow the rays of light to be bent in to a focus.

Clearly this arrangement cannot be exactly imitated in the human eye where the positions of retina and lens are permanently fixed. The same result can however be reached in another way. Take a card and prick a hole in it and hold it two or three inches in front of the eye. The hole will appear misty and blurred. The light which comes through it is spreading so rapidly that the lens is not able to bring the rays to a focus and make a sharp picture.

The rays of light are bent towards a focus but they do not cross. Now move a pin slowly between the hole and the eye so as just to touch the eyelashes. Keep the pin with its head upwards. Since an inverted picture is not formed a shadow of the pin must be cast on the retina, like the shadow of the pencil in the first part of this experiment.

Nevertheless what is seen is not this :—⊤ ; but this :—⊥.

This experiment was devised by a Frenchman named Le Cat and shows how completely the habit is formed of assuming that what affects one side of the retina is really on the other side.

The experience may be compared with that of touching a wall or the floor with a long stick. All the sensations of resistance are really in the hand and arm but they are felt to be in the end of the stick.

Accommodation.

Hold up two fingers slightly separated, or a pencil, about a foot from the face and stand before a wall. Close one eye and look at the fingers. Notice that the wall becomes indistinct and misty ; the pattern of the wall-paper can no longer be seen. Still keeping the

eye closed look next at the wall and note the change
in the appearance of the fingers whose edges become
blurred and indistinct while the wall comes into clear
view. (Cf. Fig. 35.) These changes are brought about
by changes in the shape of the lens of the eye.   In a
camera, where the lens is rigid, i.e. cannot change its
shape, different objects can be made clear in turn by
altering the distance between screen and lens.   In the
eye this is not possible and an alteration in the shape
of the lens takes place instead.   What structures have
already been described which serve to alter the shape
of the lens? (See p. 97.)

EXERCISE 49. *To observe the effect on the size of
the pupil when the amount of light entering the eye is
altered.*

Examine the eye of a person standing back to the
light. Allow him gradually to turn round until the
light falls full upon his face, keeping a careful watch
for any changes in the size of the pupil as this is done.
Make a note of the results observed. By means of a
hand glass the experiment may easily be tried upon
oneself.

The advantage of such changes in size of the iris is
easily seen. It was seen in Exercise 45, p. 90, that
when a screen, with a small hole in it, is placed in front
of a lens, the picture it casts is not reduced in size but
becomes dimmer and sharper, and on the contrary when
the hole is increased in size the picture becomes brighter
but less sharp.   The result is that in a faint light many
objects can be seen if the pupil is wide open, whereas if
the pupil were closed the field of view would be absolutely dark.

A

B

Fig. 35. Effects of accommodation
A. Eye focused on wall-paper.   B. Eye focused on pencil

EXERCISE 50.   *To illustrate the effect of alteration of the pupil upon the visibility of objects in a faint light.*

Go into a room with a candle at twilight when the light has almost gone and blow out the candle ; note the way in which objects suddenly spring into sight after a fraction of a second's interval.   When the room is entered the pupil is small or **accommodated** to the bright light of the candle ; as soon as this is put out the pupil contracts, but not instantaneously, so that faintly lighted objects do not appear for a moment or so.   This must not be taken to mean that the change in size of the iris is responsible for the whole effect.   The retina no doubt undergoes changes as well.

EXERCISE 51.   *To observe the alteration in size of the pupil of one's own eye directly.*

In Le Cat's experiment it has been shown how a shadow can be thrown upon the retina.   It is also possible to throw a shadow of the iris upon the retina. Take a card with a pin prick as before and hold an inch or two from the eye.   As before the hole appears blurred and nearly circular.   In reality what is seen is a dark shadow cast by the iris.   The light coming through the pin-hole is rapidly spreading and cannot be brought to a focus ; hence a shadow of the iris is formed just as in the case with the pin.   Without closing this eye, open the other and observe the shrinkage of the small circle of light caused by the pin-hole.   When the other eye is opened more light falls upon it and consequently its iris contracts.   Neither iris can contract alone and hence the change is also seen in the eye with the pin-hole before it.

EXERCISES.

1. Note the effect of increased and diminished light upon the eyes of any animals observable, e.g. dog, cat, rabbit, horse, sheep, etc. Record results with sketches.

2. What special features would be an advantage to the eyes of nocturnal animals? What special peculiarities are actually found, e.g. in the cat? If you can visit a menagerie compare the owl, lion, tiger, etc.

3. Carefully describe the changes which take place in the experience of sight when a brightly lit motor meets and passes you on a dark night.

4. There is a certain road recently lighted with a few street lamps rather far apart. Foot-passengers now declare that the road is darker than ever. How is this?

5. Explain what is meant by "seeing stars" after a blow.

6. Mr Pickwick's dark lantern "...was very pretty to look at but seemed to have the effect of making surrounding objects darker than before" (Ch. 39). Is this a correct description of the effect a dark lantern would have? Explain the effect so described.

7. Why does an object near the eye look larger than the same object further away?

8. On entering a lighted room after a long walk in the dark a man's eyes half close involuntarily. Suggest a possible reason beyond saying that he is dazzled.

Near point of vision.

In the last experiment it was shown how the lens of the eye alters so as to bring into clear view objects at different distances. It is now to be seen how far such **accommodation** can go.

EXERCISE 52. *To determine roughly the nearest point at which objects can be clearly seen.*

Bring this book nearer and nearer to the eye and see how soon the print begins to appear blurred. Test each eye in turn and record the nearest distance in inches at which the print is still sharp.

EXERCISE 53. *To determine the near point of vision accurately.*

Take a piece of glass tubing about 10 inches long, and two small corks. Bore holes in the corks so that

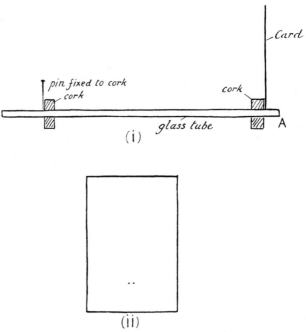

Fig. 36. Apparatus used for determining near point of vision
(i) *Section.* (ii) *Front view of Card*

one can be fixed at the end of the tube and the other slides tightly along it. One of these corks should be fixed about 1 centimetre from the end of this tube and on it may be fixed a square of card 3 inches each way. The end of the tube which bears the card should now be rested against the cheek-bone, and two small pin-pricks made in the card 1 millimetre apart just opposite the eye so as to see through them. Next a pin must be stuck into the sliding cork as far as possible from the card. The apparatus should now look like Fig. 36.

In use the end *A* is placed upon the lower rim of the eye-socket and the sliding cork, with the pin, shifted closer and closer to the eye until the pin just appears double. The distance of the pin from the near end of the tube gives the distance of near vision accurately. If a slip of centimetre paper is pushed down the tube so that its end comes exactly flush with the end of the tube, and if the pin is fixed on to the end of the cork with stamp paper or seccotine, measurements can be rapidly read off with the greatest ease.

Use this apparatus to make three measurements with each eye.

The nearest distance at which clear vision is possible is called the **near point** of vision.

Find out whether your near point changes from time to time.

Exercise 54. *To illustrate the principle of the experiment on near point of vision.*

The method in which this experiment works can be illustrated in the following way. Take a lens and arrange a candle so that an image can be thrown upon a screen of white paper. Take a piece of brown paper, make two round holes in it and place it close in front of the lens. The rays of light coming from the candle-flame are rapidly spreading as they pass through the two holes in the brown paper, but the screen can be placed in such a position that the lens will be able to bend the rays so much together that only one picture is formed. If, now, the candle is moved a little nearer, the light will be diverging at a wider angle than before and the lens will not be able to bend the two beams together to form a single picture. The picture can again be sharpened by moving the screen. In the instrument described above the pin supplies the source of light and in place of moving the screen the shape of the lens is altered. Eventually however a point is reached when the lens can change no further and the double picture remains in spite of all attempts to alter it.

Defects in the eye due to faulty shape.

The near point of distinct vision is not the same in all eyes, partly because eyes and lenses may differ in shape. Other results also follow from defects of shape, viz. the defects known as long and short sight.

EXERCISE 55. *To illustrate the result of having the back of the eye too near the lens.*

Fit up a lens, candle, and screen, as in previous experiments and move the screen gradually nearer the lens till the picture ceases to be sharp. Find out how the candle must be shifted so as to make the picture sharp again. Now suppose the lens used represents the lens of the eye in the condition when accommodation has gone as far as it can, i.e. when the lens is being used on the nearest point possible. Then it will be seen that the nearer the screen is to the lens the further from the lens must the candle be in order to give a sharp picture, or the nearer the retina is to the lens of the eye the further away is the near point of distinct vision. In the defect of the eye known as **long sight** the eye is too narrow from front to back, the retina is too near the lens.

EXERCISE 56. *To show the result of having the retina too far away from the lens.*

Repeat the last exercise but move the screen in the opposite direction. The previous result will suggest that it will now be necessary to bring the candle closer and closer to the lens in order to give a sharp picture on the screen, and experiment will prove this to be the case. Just as, in the last experiment, it was shown that a flat eye carries the near point of distinct vision away from the lens, so an unusually long eye brings the near point nearer. In other words **short sight,** or the defect of having to bring an object very close to the eye to see it clearly, is caused by having an eye too long from front to back.

Tests for long and short sight.

Suppose three persons, one long-, one short-, and one normal-sighted, stand side by side and examine a card on which are printed letters of various sizes. Now imagine the card to be taken further and further away and consider what effects are produced on the visibility of the letters to each of the three persons.

All the letters at first will seem blurred to the normal-sighted person since they will be nearer than his near point of clear vision. The same will be the case with the long-sighted person whose near point is further away. The short-sighted person may, however, be able to see the print long before either of the others.

As the letters are gradually removed they will next become clear to the normal eye, finally a point will be reached where they are clear to the long sighted.

If the type is now still further removed it will become hazy to a short-sighted eye, next to a normal eye and finally to a long-sighted eye.

It is clear, then, that a *short-sighted* person when closely examining an object will hold it close to the eye, and moreover to him distant objects will appear blurred or even be invisible.

A *long-sighted* person on the other hand can only scrutinize a small object closely by a great effort and often not at all. At the same time objects which are distant and nearly invisible to the normal eye can sometimes be seen by the long sighted.

Rough tests for long sight.

(1) Near point of distinct vision as tested in Exercises 52 and 53, p. 109, much greater than that of other people.

(2) Details in a distant view visible to you which are invisible to others.

(3) Headache or dizziness after reading fine print or doing fine needlework.

(4) Inability to count stitches in a piece of coarse linen or meshes of fine wire gauze.

(5) Blurred appearance of print after an hour's reading or more.

Rough tests for short sight.

(1) Very near point of clear vision as tested above. (This is not a certain indication ; it may show merely great power of accommodation.)

(2) Inability to see details in a distant view which are visible to most others.

(3) Power to read more easily by firelight or twilight than others.

(4) Headache and dizziness after prolonged attention to a distant blackboard in school.

EXERCISE 57. *To test for defects of sight more accurately.*

To test for defects of vision more accurately "test types" are used, i.e. a card is shown to the person to be tested, on which are printed a number of letters of various sizes. The letters of each size are known by a number which gives in metres the distance at which they should be readable by a normal individual. The person to be tested stands at such a distance that only the largest letters can be read and he reads all he can see. He then comes 1 metre nearer and repeats the test. The oculist also finds the effect of allowing him to try the test while using various lenses in front of the eye. An accurate test such as this can only be used with advantage by a specially trained observer, but the following method, described by H. Richardson[1], will serve to pick out those who require to be examined by an oculist. In this test the class are shown an oculist's test card and asked to write down all the letters they can see without straining their eyes. A short time is allowed and the results are then corrected, a mark being awarded for each line correctly copied. A fresh card is then shown and the process is repeated. By interchanging the positions of the various members of the

[1] "A Schoolmaster's Opportunities," *School Hygiene*, p. 456, 1910.

class it is possible to ensure that every individual is tested at several different distances.

Other defects in the mechanism of the eye.

Up to this point the defects of the eye which have been referred to are due to the eye being of an unusual shape, viz. too flat or too elongated from front to back. Other defects may arise from the lens itself being of a faulty shape.

EXERCISE 58. *To show what effect an irregular shaped lens has upon the picture it produces.*

Take a large round bottomed flask and fill it with water. Place a candle in front of it and on the other side a paper screen to receive the picture. Note the appearance of the picture thrown upon the screen ; sketch it. Replace the round bottomed flask by one of the oval Florence flasks in which olive oil is commonly sold. Sketch the picture produced with the flask in various positions.

These experiments should show that the effect of an irregular shaped lens is to produce a distorted picture which is too wide one way and too narrow the other.

EXERCISE 59. *To see whether the lenses of the eye cast a distorted image or not.*

Rule a series of equally thick black lines on a piece of paper, arranging them like the spokes of a wheel. (See Fig. 37.) Hold the paper at arms length and see whether each eye in turn sees all the lines of equal thickness. If any distortion, such as was produced in the last experiment, takes place the result will be that some of the lines will seem too short and broad and others too long and thin.

It is clear that it is necessary to be quite sure that the lines are really all equally dark to start with. To make sure that apparent differences are not due to real differences in the lines, turn the paper round through a right angle. If the differences are due to mistakes in

Fig. 37. (*For Exercise* 59)

vision, lines in the same position will still seem too thick ; if they are due to actual differences in the lines, the thick lines will turn round with the paper. A defect of this sort is called **astigmatism** and in some degree is not uncommon.

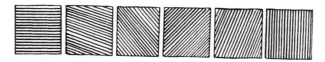

Fig. 38. How to rule squares as a test for astigmatism

(*These are intended merely as an example of the method. Other squares should be made with lines at all possible slopes*)

EXERCISE 60. *Another test for astigmatism.*

Rule a number of squares 1 inch each way and shade them with parallel lines running in different directions in each square. (See Fig. 38.) Hold the paper just so far off that the lines begin to look hazy. Now see whether all the squares appear equally darkly shaded. Test as before by turning the paper through a right angle. If differences are really due to defects in the eye a fresh square will darken as the paper is turned.

Look at your watch. Are any of the figures on the dial at all hazy or indistinct, if so which ? Turn the watch round through a right angle. Which figures are now indistinct ?

A note on the care of the eyes.

When this chapter has been worked through up to this point a note should be made of any defect of vision which has been discovered. It is not possible in a book of this kind to give full directions for testing for every kind of defect ; much less would it be practicable or wise to say exactly what should be done in any particular case. It is worth while, however, to direct attention to the following quotation (which is taken from Galton's *Descriptive List of Anthropometric Apparatus*, 1887): "There are hundreds of thousands of cases in which the eyesight has been heedlessly injured beyond repair by pure neglect ; of lop-sided growth and of stunted chest-capacity, which measurement would have manifested in their earlier stages and which could have been checked if attended to in time." It is very much hoped that the measurements suggested in the exercises of this chapter, rough as they are, may bring to view any serious defects which might otherwise pass unnoticed.

If these tests cause any suspicion that the sight is in any way imperfect, a doctor or an **oculist**, i.e. a doctor who has specialized on eyesight, should be consulted with as little delay as possible. It is not always enough to go to an **optician**, that is a man who sells spectacles. An oculist understands the human eye ; an optician understands glass lenses. A slight defect may be scarcely noticed and yet if not attended to may cause serious damage, perhaps beyond repair, and this is particularly the case with short sight. If, however, glasses are obtained from the optician as ordered by the oculist, the user is to all intents and purposes like an ordinary person while he has them on, and many defects of the eye, if taken early enough, can be cured outright.

CHAPTER XII

THE EXPERIENCE OF SIGHT

The fixation point—The blind spot—Specks in the field of view— The margins of the field—After-images—Comparison of the senses of sight and touch—Differences in sensations of sight— Difference in position—Area covered—The Galton bar—Bright- ness—Colour—The spectrum—Coloured glasses—Reflected colours—Colour mixing—The power of seeing colours—Rough tests for defective power of seeing colours—The Edridge Green classification test—The Edridge Green lantern—A caution

Apparatus and materials required.

Card. Pins. Two retort-stands. Narrow board. Sheet of cartridge paper. Coloured papers. Dark room or dark box. Glowing splinter. Cardboard. Centimetre paper. String 30 ins. Whirling table. Spectroscope. [A good pattern is the direct vision spectroscope made by John Browning, 146, Strand, London, price about 21s.] Coloured glasses or gela- tines. Small sheets of glass fixed in vertical position.

Coloured wools, silks, etc.   Edridge Green colour tests.
Holmgren's worsteds.   Mathematical Tables of Trigo-
nometrical Functions.

In this chapter we shall go on to consider what
experiences are supplied by the eyes quite apart from
the manner of their production.

EXERCISE 61.   *To show that in each eye one part*
of the field of view is more clearly seen than the rest.

The picture which is seen with one eye is called the
field of vision from that eye.   Take page 121 of this
book, find letter "a" in the word "any" near the middle
of the page.   Close one eye and look steadily at this
letter with the other.   Find out how many letters above
and below and to the right and left of this can be seen
clearly without moving the eye.   It is well in this
exercise to get a companion to watch your eye carefully and make sure that it does not move involuntarily.

The result shows that only a very small part of the
field of vision is clear at any one time.   Experiments
have shown that it is that part which is seen by the
yellow spot of the retina (see p. 97).   This point in the
field of view is called the **fixation point**.   In ordinary
language it is called the point looked at.

EXERCISE 62.   *To show that in each field of view*
one point is not seen at all.

Use the right eye and close the left.   Place the two
thumbs side by side with the nails uppermost on the
edge of the table.   Keeping the right eye fixed upon
the left thumb, gradually move the right thumb to the
right.   After moving a few inches it disappears and
after moving a few inches further reappears.   Test the
other eye in the same way.

This experiment shows that in each field of view there is a portion which is unseen and in each eye a spot which is unable to see. This spot in the eye is called the **blind spot** and has been shown to correspond to the part of the retina where the optic nerve enters it. On which side of the optic nerve is the yellow spot? Verify your answer by referring to your drawing of the rear half of the eye.

EXERCISE 63.   *To show the presence of specks in the field of view.*

Prick a hole in a piece of card and look through it at a clear sky. Small specks and curved threads can usually be seen floating across the field. They are due to small bodies in the vitreous humour and are named **muscae volitantes**, which is the Latin for fluttering flies. They are a perfectly natural appearance and can usually be seen in any healthy eye. Sometimes they may be seen without any special apparatus, when one happens to be looking at a sheet of white paper, an expanse of snow, or a clear sky, and they are nearly always seen on looking down the tube of a microscope.

If instead of specks floating across the field there is a black spot blocking out part of it or fine threads in a fixed position, the eye should be examined by some competent person.

EXERCISE 64.   *To test the margins of the field of view for the power of seeing colours.*

Fix a pin vertically into the centre of a sheet of cartridge paper and arrange two retort-stands with a narrow board between them to serve as a head-rest. (See Fig. 39.) By this means it can be arranged to have the head supported in such a position that the

eye to be tested looks vertically down upon the head
of the pin which serves as a fixation point.   The advantage of having a pin and not a mere dot on the paper
is that with a pin one's head can be kept in the same
position with more certainty than with a spot.   Now get
a companion to slip into the field of view, from the
margin, pieces of paper of various colours.   These may
be about 1 cm. square and can be stuck on slips of
white card 5 cm. by 3 cm. for convenience of handling.
Describe carefully all the changes in the appearance
of the coloured squares as they pass from the circum-

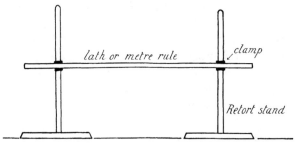

Fig. 39.   A simple head-rest

ference toward the centre.   How near the fixation point
do the colours become certainly recognisable?   The
answer clearly depends upon the distance of the eye
from the paper.   If there is any doubt about this, test
it by experiment.   It is accordingly not enough to
measure the distance from the point where the colour
is first seen to the fixation point.   Instead of this it is
usual to measure the angle between two lines: first, that
from the eye to fixation point, second, that from the eye
to the place where the colour is first seen.   These lines
are shown as AC and AB in the diagram (Fig. 40).

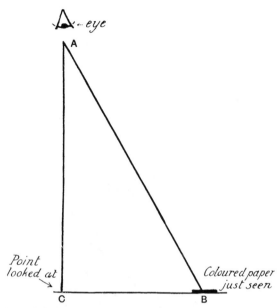

Fig. 40. Method of measuring positions where colours are just seen

Table of Natural Tangents

Angle at A	Tangent $= CB/CA$
0°	·0000
5°	·0875
10°	·1763
15°	·2679
20°	·3640
25°	·4663
30°	·5774
35°	·7002
40°	·8391
45°	1·0000

See "Mathematical Tables," published for the Board of Education by Wyman and Sons, price 5s. per 100.

To measure this angle get your teacher to show you how to use a table of natural tangents. The easiest way to use this is to make AC exactly 10 inches and measure BC carefully in inches. In the table two columns will be seen. One of these is the result of dividing CB by CA. The other is the size of the angle at A.

EXERCISE 65.   *To see the blood vessels of the retina.*

Hold a lighted candle to the outer corner of the eye, as close to the face as is possible without burning it.

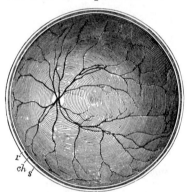

Fig. 41.   Blood vessels of Retina
r, *Retina*   ch, *Choroid*   s, *Sclerotic*

Close the other eye and look steadily into a dark room or dark box. Raise and lower the candle gently an inch or two at a time. The field of view will appear a dull red colour and on it, like the pattern on shot silk, will be seen a number of branching lines. (See Fig. 41.) Their general arrangement is like that of river systems on a map and they seem all to run together toward a point roughly in the centre of the field.

Portions of this pattern are sometimes to be seen in the ordinary field of view without special lighting.

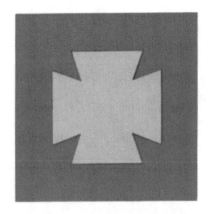

Fig. 42

EXERCISE 66. *To show that a sensation lasts after the stimulus which causes it has ceased.*

Take a splint of wood and set fire to one end. After it has been burning for a few moments blow it out. While the end is still glowing wave it rapidly about in several directions and note the appearance of the glowing end. It is seen as a continuous line of light owing to the fact that the sensation in the retina takes a short time to die away. The sensation which persists in this way is called the **positive after-image**.

EXERCISE 67. *To study the changes which take place in a sensation as it dies away.*

Look steadily at the bright flame of a lamp or better still at the wires of an incandescent electric lamp for 30 secs. or a minute and then at a piece of white paper. Notice that the positive after-image only lasts a very brief fraction of a second; often, in fact, vanishes before the eye reaches the paper, and it is then replaced by a dark blot the same size and shape as the object first looked at. This is called the **negative after-image**.

Obtain a good after-image of a white light and carefully record the various colours seen as it fades.

See whether an after-image formed by looking at an object with one eye only, can afterwards be seen with the other eye.

Look steadily at Fig. 42 in a good light and then look at a sheet of white paper or at the ceiling. Describe what you see.

A comparison of the senses of sight and touch.

Recalling the results of the experiments on sight up to this point, it is seen that the experience of sight is very different in different parts of the field of view.   In this point we have a marked difference from the sense of touch.   Although touch does vary somewhat from place to place, it does not do so in quite the same way as sight does.   A prick with a pin feels much the same whether it is in the hand or on the leg.   A pin looked at near the fixation point on the other hand looks very different indeed from the same pin when it is close to the margin of the field of vision.   We have seen that there is only one small spot where objects are clearly seen at all, that there is a spot not far from this where *nothing* is seen at all and that round the margin of the field no colour is visible.   Added to this a large number of the objects seen are completely out of focus and, if the eyes are shifting about, everything seen is mixed with the fading after-image of the last thing looked at. Finally it is quite possible that in many cases floating specks of dust, etc. in the eye may still further interfere with the picture.   It is not surprising that a certain well-known writer on the subject should have said "If an optician wanted to sell me an instrument which had all these defects I should decline to take it off his hands on purely optical grounds."

These difficulties are overcome by the unconscious use of the imagination.   It is not until definite experiments, of the kind described above, are made that most people are aware of these optical defects at all.

When an object has once been seen at the fixation point, all that is seen to be true of it there is still supposed to be true of it when it is seen in the margin of the field.   And not only is it supposed to be true but the person in question thinks he sees it as before.   The eyes in fact roam about so freely that all the objects seen in the field at one time seem to be seen by this means with practically the same detail as regards colour, clearness, etc. as they are at the fixation point.

We now leave the question of the differences due to the different parts of the field of vision and consider in what ways various sensations of sight may differ from one another.

EXERCISE 68. *To illustrate the way in which sight sensations differ from one another.*

Take a piece of white paper one inch square and without moving the eye shift the paper a few inches. It is now seen in a different place. Sight sensations are therefore similar to touch sensations in that they differ according to the **place** where the sensation occurs.

Now place alongside the first square a second square of equal size. What is now seen differs from what was at first seen in that a larger area is covered. Thus sight sensations are similar to touch sensations in that they differ according to **size.**

Keeping the eye fixed on the two squares, draw down the blinds. The only difference now is that the squares of paper look less **bright**. The only differences in touch sensations at all like this are differences in pressure and in temperature, and even here the correspondence is by no means close.

Repeat the experiment once more, this time examining the cards by a coloured light or replacing the white cards by coloured ones. This will show a change in sensation as regards **colour.** There is no change in touch sensations at all like this except perhaps temperature.

Thus sight sensations differ from one another as regards

     1.   Size.
     2.   Position.
     3.   Brightness.
     4.   Colour.

Difference in position.

In measuring the accuracy with which the positions of touches could be judged, two points were used and the distance was found at which they could not be separately distinguished. A similar method can be used here.

EXERCISE 69. *To find how far apart two points must be to be just seen as two.*

A difficulty occurs here exactly like that of Ex. 64 (p. 121). There the distance from the fixation point at which colours could just be distinguished was measured. It was, however, found convenient to measure not an actual distance but a certain angle. In the same way, when we want to know how near together two points must be to be just distinguishable, it is usual to measure, not the actual distance, but the angle made by two lines meeting at the eye one drawn from the centre of each point.

Take a piece of white cardboard and make two squares on it each 1 mm. square and exactly 1 mm. apart. Place the card in a good light and gradually move away from it till a point is reached at which the two dots appear to fuse into one. Measure the distance from the eye to the card in millimetres.

To find the angle in minutes made by the two lines where they reach the eye a table can be used. Get your teacher to explain to you how to use a "table of circular measure" which may be found in any book of mathematical tables.

Table of Circular Measure.

Angle in minutes	Circular measure $= \dfrac{\text{arc}}{\text{radius}}$
0′	·00000
1′	·00029 +
2′	·00058 +
3′	·00087 +
4′	·00116 +
5′	·00145 +

See *Chambers' Mathematical Tables,* published by Chambers, page 251.

For these very small angles the circular measure is practically the same as the natural tangent already explained. For these small angles the difference may be neglected and either table may be used.

A question arises here as to whether a person who ordinarily uses spectacles should use them when making the test. The answer depends upon what it is desired to test. If it is required to test merely the power of the retina to distinguish the points apart, then clearly spectacles ought to be used so as to overcome any defect arising merely from incorrect shape of the eyeball. Make five tests using each eye separately.

Area covered.

Things seen are noticeably different in size, or in other words, sensations of sight differ in the areas over which they can be felt. The accuracy with which the sizes of objects can be distinguished can be measured by an application of some of the methods that have already been used in testing other senses.

The following experiment is very rough. Its principal use is to show some dangers to be guarded against.

EXERCISE 70. *To see how nearly a line can be drawn equal to a given line.*

Take a sheet of white paper and draw a horizontal line near the top of it 5 cm. long. Look steadily at the line with the eye to be tested for three seconds, cover with a card and draw lower down on the same paper another line which seems to be of the same length. Cover this line, look again at the first and repeat. After ten trials measure all the lines drawn. Calculate the mean and the mean error as in the exercise on the distances moved by the hand (Ex. 22, p. 52). We have now to consider whether these results really measure the accuracy with which the lengths of lines can be distinguished by eye. There are at least three obvious objections. In the first place no attempt has been made to distinguish between the length of the line as judged by the space it covers in the field of view when looked at, and its length as judged by "running the eye along" it.

In the second place it is quite likely that the hand has not correctly carried out the movement which was intended, indeed it will often be found that, if india-rubber is allowed, a much exacter result can be obtained.

Finally at the time when the line is drawn the standard line is no longer visible, and errors depending on memory are thus introduced.

The Galton bar.

To overcome these difficulties an instrument known as the Galton bar can be used. It is named after its inventor who has already been referred to several times in this book. It consists of a bar with a transverse scratch in the centre. At each side is a slider which can be moved easily along the bar and fixed in any desired position.

In use one of the sliders is set at a fixed distance from the middle line by the experimenter, and the person to be tested sets the other at a distance which seems to him to be the same.

EXERCISE 71. *To make and use a rough Galton bar.*

Take a strip of card 1 in. wide and about 20 ins. long. Stick on one side of it a strip of centimetre paper; this is to be the back of it. On the front of it and exactly opposite the central division of the centimetre paper rule a transverse line with a sharp pencil so as to divide the strip into two equal parts. Make two sliders out of strips of paper folded round the card strip and gummed by their edges.

A figure of this instrument, and also of another which any one can make for himself, is shown below (Fig. 43). The rough card here described gives quite satisfactory results. To use it the slider on one side is set at a fixed distance from the middle point of the bar, e.g. 5 cm. The bar is then handed to the person to be tested who sets the other slider at what seems to him the same distance from the middle line, without, of course, referring to the graduations on the back.

Before using it, it should be settled whether eye movements are to be allowed or not. If the judgments

are to be made without moving the eye at all the
experimenter should watch very carefully to see that
this is properly attended to.

In making the tests it will be seen that there are
four ways in which each may be made. The standard
line may be to right or left of the other, and the line
to be altered may be too small at first or too large at
first. In making the tests an equal number should be

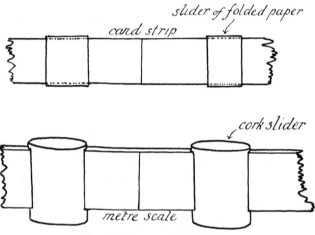

Fig. 43.   Two rough Galton bars

made by each method so that differences due to these
various causes may balance one another as far as
possible.

Use this method for each eye separately, finding in
each case the mean of the whole series, and also the
mean of all individual differences from the mean (cf.
method of Ex. 22, p. 52).

Try and devise methods of finding out whether any
difference is produced by allowing the eye to move

and also see whether the results can be arranged so as to show what effect is really produced by the position of the standard line to right or left of the other line, and also by the original size of the line to be altered.

Estimation of tenths.

If a line is drawn on the blackboard about 1 metre long and one end of it called 0 and the other end called 10, then if any chalk mark is made between 0 and 10 it will generally be possible to guess its position to the nearest unit, calling it 5 if midway between 0 and 10, or 9 if close to 10, or 3 if about one-third of the way along. For practice it is convenient to have the exact tenths marked on the blackboard in some way invisible to the class, but visible to the teacher for reference. Quick and accurate estimation is soon learnt, and after this it is possible to use a ruler graduated in tenths of an inch and by estimation of tenths of tenths to measure accurately to the hundredth of an inch.

Brightness.

It has already been seen (Exs. 66 and 67, p. 125) that when an object has been observed and then removed the sensation does not die away at once but fades slowly. If a second object is looked at before the first sensation has faded, the two cannot be distinguished from one another. If, for instance, a pencil is very rapidly waved backwards and forwards over this page, a sort of fan-shaped haze is produced by the successive positions of the pencil, while at the same time the type can be seen through it. Use is made of this circumstance in making experiments on brightness. A disc of card is cut out and a black mark made upon it. The disc is then

rapidly spun round and the black mark is seen as if
it were a grey band.

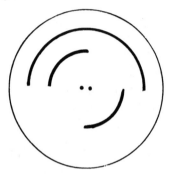

Fig. 44.
Card disc for experiments on brightness

EXERCISE 72.   *To show the way in which sensations
of black and white can be combined if they follow one
another rapidly.*

Take a card six inches in diameter and mark out
two rings on it.   In each case arrange to have half the

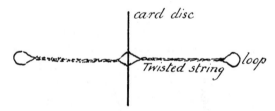

Fig. 45.   Method of spinning card disc

ring white and half black, but in one case have the black
half in a continuous semicircle and in the other in two
black quarter circles with white quarters between them
(see Fig. 44).

Make two holes through the disc and thread it on a loop of string and spin it rapidly by this means (see Fig. 45). The disc spins regularly first one way and then the other. Compare the tint of the two grey rings when the disc is spinning at its fastest. This experiment illustrates the fact that when white and black sensations follow one another rapidly, a grey is produced which is similar in tint to the colour which would be obtained by mixing white and black in proportion to the times for which the sensations last.

This suggests a method of measuring changes in brightness just large enough to be noticeable.

EXERCISE 73. *To see how much the brightness of a piece of white paper must be diminished to become noticeably greyer*[1].

Cut out a disc of white card or Bristol board about 10 cm. radius. Along one radius make a row of black marks 1 mm. sq. in Indian ink. Fix the card on a whirling table such as is used in physical experiments ; turn it rapidly so that grey bands are produced, and find out which is the faintest ring which can be detected. Take care to carry out the experiment in a good light but not direct sunlight, and also avoid moving the eyes about rapidly (see Fig. 46).

Suppose it is found that the faintest ring visible is 20 mm. from the centre of the circle. Calculation shows that the circumference of this ring was $2 \times 3\cdot14 \times 20^2$. Multiplying out this gives 125·6 mm. Of this 125·6 mm.

---

[1] The method here outlined is due to Sanford, *Experimental Psychology*, p. 138.

[2] From the formula

$$\text{circ.} = \pi \times \text{diam.,}$$
$$\pi = 3\cdot14.$$

one was black and the remaining 124·6 were the "white"
of the paper. When the disc was rapidly revolved this
grey band must have been $\dfrac{124\cdot6}{125\cdot6}$ as bright as the "white"
of the disc ; in other words the " white " of the disc had
to be reduced $\dfrac{1}{125\cdot6}$ of its original brightness before the
difference was noticeable.

Question.—Is it correct to put the word white in quotation marks
in this passage ?

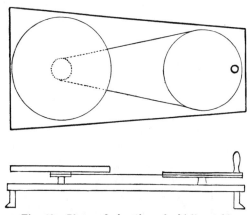

Fig. 46.   Plan and elevation of whirling table

Colour.   Spectrum Colours.   The Spectroscope.

It has been shown on p. 89 that when light coming
from an object passes through a prism or a block of glass
the rays are diverted from their original course.   Light
of some colours is bent much more than that of others.
Advantage is taken of this fact in the instrument known
as the spectroscope.   In this a very narrow beam, obtained by passing light through a narrow slit, is made

to go through a series of prisms so that a much greater amount of bending takes place than when only a single prism is used.

EXERCISE 74. *Use a spectroscope to examine the appearance of the slit when seen through the prisms.*

Note that the light slit is spread out into a long band, note further that the colour of this band is not the same all along, but is coloured differently in different parts. The white light, as such, has in fact disappeared and in its place a number of colours have made their appearance. This coloured band is called the **spectrum** and the colours seen in it are named **spectrum colours**.

Make a neat sketch of the spectrum as you see it and mark, if you can, the boundaries of the different colours. Label these with names which you think describe them. If simple names such as blue, bluish green, etc. cannot be found for them, compare them to common objects, e.g. colour of a geranium, colour of grass, etc. If possible, paint or crayon your sketch.

When the results are complete and not before compare the results obtained by others and make a note of any difference of opinion.

Do either of these lists describe accurately what you see?

(*a*) Red, Green, Violet.

(*b*) Red, Orange, Yellow, Green, Blue, Indigo, Violet.

If not, how would you correct these lists to give a truer account? Do the different colours occupy equal space? Do the colours change into each other very gradually or are their boundaries fairly sharp?

Part of the explanation of the facts recorded here

is that white light is in reality a mixture of all the colours seen through the spectroscope. The prisms of this instrument bend light of some colours more than that of others, and consequently the light coming through the slit is broken up and arranged into a many-coloured band. Try to decide which of the colours is bent most and which least.

EXERCISE 75. *To find what is the effect of placing coloured glasses in front of the slit of the spectroscope.*

Use coloured glasses or coloured slips of gelatine such as is used for Christmas crackers.

Select a piece of blue glass and hold in front of the slit of the spectroscope. It will be found that the apparently blue light which comes through the glass is still split up by the spectroscope, and as well as the pure blue light a variety of other colours can also be detected. These other colours will appear as coloured bands in their usual places in the spectrum.

This experiment illustrates the fact that coloured glasses really act as a kind of filter and absorb nearly all the white light. Thus blue glass absorbs or filters out all but the blue and small quantities of other colours ; red glass filters out all but the red and small quantities of other colours, and so on.

Question.—What happens if you look through two pieces of glass one over the other, one of them pure red and the other pure green ? Try and prophesy the result and then test by experiment.

Reflected colours.

Many objects are seen simply because of the light they reflect. What has just been said about transparent colours will suggest the reason why opaque objects are sometimes coloured. Most of the white light which falls upon such objects is not reflected at all but absorbed by them, and this action of absorption has more effect upon some colours than others. As a consequence the light which is reflected is often coloured. It is really white light from which the colours which the object has absorbed have been subtracted.

EXERCISE 76. *To test the purity of the colours given by opaque objects.*

Use the spectroscope as in the experiment with coloured glasses to see whether the light reflected from various coloured papers is pure or not.

Allow white light from a north window to be reflected into the slit of a spectroscope by a slip of paper placed at an angle in front of it. It will be found that with most coloured papers obtainable the colour reflected is much less pure than that transmitted by coloured glasses or gelatines[1].

Colour mixing.

In the last two experiments it has been seen that when matters are so arranged that varying quantities of light of different colours fall upon the eye it often happens that only one of the colours can be seen. In both these cases however it has happened that one colour has been present in much greater proportion than the others. The effect is perhaps somewhat similar to that of listening to a band in the distance. Sounds from all the instruments reach the ear, but it seems to the listener as if only the drum were playing. It remains to be seen whether the comparison will hold when the colours are present in nearly equal proportions.

---

[1] An extensive series of coloured papers is manufactured by the Educational Supply Association, Holborn Viaduct, London, E.C.

In a fife and drum band close at hand the instruments can be detected separately.

If red and green are mixed in equal proportions, can they be separately distinguished as such?

EXERCISE 77. *To see what is the effect of mixing red and green light.*

Fix a plate of glass in a vertical position and put two squares of coloured paper flat on the table, one on each side of the glass. With a little manipulation a point can be found where the two papers are in such a position that the reflection of one from the surface of the glass just covers the other as seen through it. What colour is seen as a result? What result is produced by exchanging the positions of the two pieces of paper? In the same way record the result of mixing the colours of any other bright coloured papers which can be obtained.

The power of seeing colours.

This is not a suitable place for a discussion of the various theories which have been suggested to explain the differences in the way in which colour is differently seen by different persons. Those interested in the subject should refer to the list of works quoted in the appendix.

Broadly speaking the differences between various persons arise from the fact that certain colours are, to some persons, either invisible or indistinguishable from other colours. When it is remembered that many colours, e.g. coloured glasses and most opaque colours (see p. 139), are really mixtures it will be seen that any defect of this kind will have the most far-reaching effects.

Rough tests for power of seeing colours.

The surest way of finding out whether your power of seeing colours is satisfactory is to compare yourself in as many ways as possible with those who are known to be perfect in that respect. The following questions are intended to suggest very rough tests.

(1) In marking out and naming the colours of the spectrum do you agree with others as to their boundaries and names ?

(2) Do the names given to colours by others seem to you satisfactory or do you find that they give different names to what seem to you shades of the same colour ?

(3) In copying a painting do the colour matches you make satisfy other people ?

(4) Carefully compare the tints seen in a sunset, a coloured fruit, e.g. an apple, a landscape, with those which others report.

(5) Do you admire the colours of stamps, of birds' eggs, of butterflies, of the kaleidoscope ?

EXERCISE 78. *To test the power of seeing colours by sorting coloured objects.*

Make a collection of as many coloured objects as can be obtained and let them be as various as possible—wools, silks, coloured papers, slips of coloured glass or gelatine, etc. There should, if possible, be several different shades of each colour.

Sort these into groups, putting into the same group all those objects which seem the same colour and appear only to differ in shade. Do not compare results with anyone else until the whole is complete. Then make a careful record of all differences between your classification and that of a person with ordinary power of seeing colour.

The Edridge Green Classification Test.

This is a test arranged on similar lines to the last but of a much more accurate character. It consists of a series of standard colours and a very large number of other coloured objects, e.g. silks, wools, coloured glasses, etc. The person to be tested is given one of the test colours and asked to pick out all objects of the same colour. The colours are so arranged as to give the greatest possible difficulty to a person who is colour-blind and yet be quite easy to those who have normal sight.

This test can be obtained from E. B. Meyrowitz, 1a, Old Bond Street, London, W., price £1. 1s. 0d. A simpler form is also made under the name of the Pocket Classification Test, price 12s. 6d.

The Edridge Green Lantern.

It is possible that a person whose sight is imperfect might still, with sufficient training, pass the classification tests. To get over the difficulty thus caused the Edridge Green Lantern has been invented. This consists of a lamp in front of which a series of variously coloured glasses can be arranged somewhat in the same way as the red and green glass slides in front of a signal lamp. In addition to altering the colour the character of the light can also be altered by ground glasses, ribbed glasses, etc., which can be used either together with the colours or alone.

The person to be tested is seated not less than fifteen feet from the lantern, and a large series of combinations of colours are then shown. The order in which they are taken is quite haphazard and as many available colours as can be arranged are tried. After each colour is shown, the person tested writes down the name of it. The papers are then collected and corrected by the examiner.

This piece of apparatus can be obtained from the same address as the last-mentioned, price £5. 12s. 0d.

A caution.

In considering these results it is necessary to bear in mind that mistakes may arise from ignorance of the names of colours. Any mistake made should, however, be looked into. If a mistake is made the person tested should not be informed of it at once; it is better to repeat the test later in the series. Any mistake which cannot be *certainly* traced to an ignorance of colour names should be looked into and the person making it tested for colour-blindness by a competent person.

PROBLEMS AND EXERCISES.

1. How is it that a dropped collar stud cannot always be found at once, even though it may be in the open and not covered by anything?

2. Describe as exactly as possible what you do when you "look for" anything. What process takes place when anything is "found"?

3. Do after-images occur in any other senses than that of sight?

4. The after-images of coloured objects often change their colour according to the colour of the background; why is this?

5. A moving wheel often looks simply like a blur of spokes; why is this?

6. Light passes through pure red and then through pure green glass; what colour reaches the eye?

7. Why does an instantaneous photograph of a breaking wave look like marble ?

8. After looking at a bright sunset and then looking at a blue sky you can see several purple suns scattered about. Explain this.

9. Why does the large hand of a watch seem stationary ?

10. A green leaf when submerged in a brown solution of iodine looks black. When taken out of the solution it is seen to be still green. How is this ?

11. A bird is seen flying at a distance. It is often many seconds before one can decide whether it is approaching or receding ; why is this ?

12. What faculties of the eye are called into play (*a*) in hunting for mushrooms, (*b*) in catching a cricket ball ?

13. If you first look at an unshaded electric lamp and then go into a dark room you see a ghost. Why is this ?

14. Examine with a microscope (*a*) a three-colour process print, (*b*) a Lumière process lantern slide. How many different colours can you see in each with the microscope ?

15. A spectrum is thrown on to the lantern screen by means of a limelight and carbon bisulphide prism. Make a diagram like this

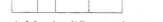

to show the areas occupied by the different colours you see (*a*) with a narrow slit in the lantern, (*b*) with a broad slit.

16. In a chemist's shop window there are two large bottles, one filled with yellow and the other with blue solution. If I look at the yellow one through the blue it appears grey. The blue one seen through the yellow seems black. Explain these facts.

17. Some electric lights really give a very rapid series of very short flashes. The light seems continuous, but if I wave about a piece of paper I can see the paper in several places at once. Why is this ?

18. Rule 10 pencil lines 15 cm. long. Try and find the middle point of each by eye. Measure your results and find by how much each attempt is incorrect.

19. How many stars can you see in the constellation of the Pleiades which rises in the N.E. on autumn evenings? Draw what you see. Then use an opera glass to find whether your drawing is correct.

20. Draw the constellation called the Great Bear. Letter the stars in order *a*, *β*, *γ*, *δ*, *ε*, *ζ*, *η*. Now number them in order of brightness, calling the brightest 1, the others 2, 3, 4, 5, 6, 7.

21. *Colour Tests.* The following colour tests are described at length in "A Schoolmaster's Opportunities," *School Hygiene*, No. 8, Vol. I, Aug. 1910, p. 456.

About one or two dozen different coloured wools are obtained including several green and pink shades. These are unwound and stranded together into one multicoloured rope. This rope is then cut with scissors into short sections about two inches long. Each pupil is given two sections and asked to arrange the colours in pairs.

22. About one or two dozen sheets of differently coloured tissue papers are laid in a pile. From this pile strips two inches wide are cut and these are again cut at right angles so as to produce little heaps of two-inch squares. Each pupil is given two heaps—shuffled—and asked to arrange the papers and pin them together in pairs.

CHAPTER XIII

ACTION

The muscles in action—Kinds of action—Reaction times—Rapidly repeated actions—Habitual actions

Apparatus and materials required.

Metronome or stop-watch. Retort stand. Square of linoleum. Penny. Pencils. Paper. Squared paper.

Use of muscles in action.

In Chaps. V and VI mention has been made of the

use of muscles in moving the limbs and of the information as to the shapes and sizes of objects obtained by their means.

The muscles can also be used to carry out a great variety of intentional movements of other kinds. Some of these are here considered.

EXERCISE 79. *To illustrate various kinds of action.*

Carefully consider and describe exactly what takes place when a letter is written. The actual action itself is in many cases only a matter of a few minutes, but a great deal goes on before pen touches paper at all. Thus one usually thinks first of all of what has to be said and to whom ; then follows the consideration of whether a letter or a post-card shall be used ; then of the words to be used and their arrangement ; finally the letter is worked out sentence by sentence and the words thought of represented by muscular action upon the paper.

The following are two careful descriptions of what takes place in performing two simple actions, viz. (1) helping oneself to salt at a meal, and (2) putting a shovelful of coal on the fire. Try and describe these actions for yourself before reading the descriptions given in small print below.

1. **Helping oneself to salt at a meal.** The salt-cellar at present being used by someone else catches my eye. || I ask for the salt to be handed to me. Using the muscles of my arm I help myself to salt.

2. **Putting coal on the fire.** Sitting writing back to the fire the room begins to feel cold. I notice this but still go on writing. At last it gets so cold that I stop writing || and look towards the fire. I see it to be dull red and dying. I get up, cross the room and put coal on.

The mark ‖ shows the point where muscular action begins in each case.

Describe in the same way as the above what happens when

3. You kill a wasp which buzzes uncomfortably near your face.
· 4. You play a ball at cricket.
5. You take a header into the swimming bath.
6. You accept by letter an invitation to tea.
7. You take out your books for an arithmetic lesson.

The value of these exercises lies entirely in the information they give concerning what passes in the mind before the action of the muscles takes place at all. Very careful attention should be paid to this, otherwise the experiment is without point. This exercise will show that in many cases it is difficult to give a clear explanation of the train of thought giving rise to the action. Some actions take place with still less thought; thus if, by accident, the hand is plunged into hot water, it is snatched out purely mechanically with no reflection whatever.

A still more mechanical action is that of closing the iris in a bright light (see p. 106). In fact as we have seen this action is so purely mechanical that the owner of the eye only becomes aware of it by using some special method of experiment. Such actions as these are called **reflex actions.**

Reaction times.

The time taken by different persons to become aware of a stimulus and make some simple movement in response to it varies considerably. It may cause some surprise to find that it is a time capable of being measured at all.

EXERCISE 80. *To measure roughly the average time taken by the brain to act in response to a stimulus.*

Arrange the class in a chain, the right hand of each member resting on the left hand of the next. In addition to the chain thus formed there should also be a starter who holds in his hand a stop-watch. When all is prepared, and all eyes closed, the starter starts the stopwatch and at the same instant squeezes the left hand of the first member of the chain. Immediately on becoming aware of the stimulus No. 2 passes it on to No. 3, who communicates the touch to No. 4 and so it passes through the whole chain. Directly the stimulus has started, the starter at once places the stop-watch in the right hand of the last member of the series. As soon as he feels the touch he stops the watch which thus records the time taken by all the members of the class to feel and answer to a touch stimulus. If the total time is now divided by the number of individuals, the result gives the average time taken by one member of the class to feel and reply to a touch-stimulus.

Different classes will give different results and the same class will usually show rapid improvement by practice. Several trials should be made, as, if a single boy is slow or inattentive, the time may be greatly increased.

EXERCISE 81. *To measure the time taken by an individual to answer to a seen signal.*

The method which is here described is due to Mr H. Richardson[1]. Take a retort stand and ring, and fix the ring about five inches above the stand. The

---

[1] "A Schoolmaster's Opportunities," *School Hygiene*, Vol. I, Aug. 1910, p. 456.

experimenter now takes a penny and the person experimented upon a thick card, or piece of linoleum, or one of the square asbestos pads used with tripod stands.

The penny is then let fall from the level of the ring and the person tested watches to see it begin to fall. As soon as he sees it actually falling he snatches away the card. The experimenter notes whether he has been able to withdraw the card or not before the penny clangs upon the iron base of the retort stand. If he has not been able to withdraw the card in time, the natural conclusion is that the time which the penny has taken to fall from ring to stand is less than the time taken by the person tested to respond to the sight of the penny. By gradually raising the ring a point can be found at which the pad fails to be removed in two out of three or, if greater accuracy is required, in four out of five trials.

The time that the penny takes to fall can be easily calculated by the use of the formula

$$t = \sqrt{\left(\frac{s}{192}\right)}.$$

In this formula t stands for the required time in seconds and s for the distance in inches through which the penny has fallen.

In an actual experiment it was found that the pad could just be snatched away when the penny had to fall 11·5 inches. Putting this value in the above formula, we obtain

$$t = \sqrt{\left(\frac{11·5}{192}\right)} \text{ sec.};$$

simplifying

$$t = \sqrt{·059} \text{ sec.}$$

Extracting the square root

$$t = \cdot24 \text{ sec.}$$

The average result usually obtained for tests of this kind of reaction when much more delicate apparatus is used is ·27 sec.

Use this method to test your right and left hands, making several tests with each hand.

Find out whether any difference is found in the result when all the attention is directed to the action of removing the card instead of being entirely directed to watching for the penny.

The author has found it an advantage to fix a piece of curved brown paper into the ring so as to conceal the experimenter's hand, and also to give a warning "*now*" half a second or a second before the penny is released.

Rapidly repeated actions.

Another useful measurement which has often been made is the time taken in repeating the same simple action, e.g. tapping, with the greatest possible rapidity.

EXERCISE 82. *To see how many taps can be made on a piece of paper in a given time.*

Take a strip of paper and a sharp pencil. Hold the pencil vertically in the right hand and the paper by one end in the left. At the given signal "*go*" start making dots on the paper as rapidly as possible, at the same time slowly drawing the paper along with the left hand. Give the signal "*stop*" exactly 20 secs. after the signal to start. Count the number of dots which have been made. Repeat the experiment 10 times, allowing intervals so as to avoid fatigue. Count the mean number of dots made in 20 secs.   Repeat the experiment, using the left hand.

Habitual actions.

Many actions which when performed for the first time require the closest attention gradually become

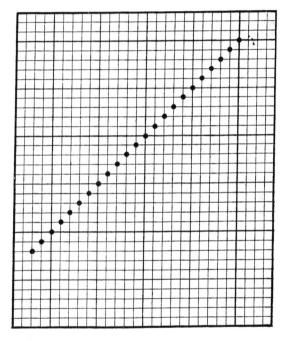

Fig. 47.   Method of making dots in Ex. 83

almost unconscious or reflex.   A familiar example is the case of walking.   At first this is a matter of the greatest difficulty, but as time goes on it becomes so purely mechanical that the walker is scarcely conscious of the actual details of the action at all.   For this reason, as has been already noticed, great difficulty is often found in correctly describing the steps in the progress of many quite familiar actions. Such actions are called **habitual actions** or, more loosely, simply **habits**.

EXERCISE 83. *To study the growth of a simple habitual action.*

Take a sheet of squared paper ruled in inches and tenths and lay it on the desk. Then take a sharp-pointed lead pencil in the left hand and make a row of dots at the intersections of the lines, running obliquely down and to the left (see Fig. 47).

Make quite certain that it is clearly understood what is to be done before the experiment is tried.

Now try and see how many dots can be made in 20 seconds. Repeat the experiment 20 times. Here is a set of results obtained from such an experiment :

No. of exp. in series	1	2	3	4	5	6	7	8	9	10	11	12	13	14	15	16	17	18
No. of dots made	12	15	15	16	17	18	20	20	23	21	20	21	21	22	24	24	23	24

The results are much more easily understood when arranged in the form of a graph (see Fig. 48).

The continuous line joining the dots surrounded by circles shows the actual results obtained. The dotted line is a smooth curve drawn through the points with a whalebone ruler. It is at once seen that in forming this habit there has been a rapid improvement at first and a slower improvement later. Two convenient measures can be made. (1) We may note the °/₀ amount of improvement which takes place in 20 trials, in this case 40 %, or (2) We may continue the trials till no further improvement can be observed and note how

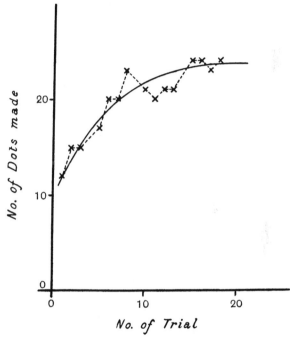

Fig. 48. Graph to show improvement in number of dots made in successive trials

many trials are required before this takes place. Clearly the first measurement is a more reliable test than the second. Make a careful test of yourself as here directed and compare the amount of improvement with that of others working under exactly similar conditions.

Other experiments on habit formation.

Other subjects which have been proposed for the study of the development of habitual actions fall into two classes : (1) Those in which the rate of performing a simple action is tested, (2) Those in which the accuracy of the performance is measured. Most tests are capable of being used in both ways, e.g. in the one here proposed we might have measured the distance of the centre of each dot from the intersection of the lines upon which it was supposed to be placed, which would then supply a measure of accuracy.

Two of the more important suggested methods of measuring habit are mentioned below; for details reference should be made to the works and especially the periodicals referred to in the appendix[1].

(1) The drawing of stars and other geometrical figures as seen by their reflection in a mirror.

(2) The substitution of various arbitrary symbols for some or all of the letters in a passage of prose.

Measurable habitual actions are continually cropping up in the teaching of every school subject, most notably in Science, in Woodwork and in Drawing. The use of the balance, the metre scale, the pipette, the measuring jar, and the thermometer, are all of them actions which as time proceeds become habitual.

It may be noted as a not unsuitable conclusion to this chapter that the important time in the formation of a habit is clearly in its early stages when the improvement takes place most rapidly. Hence in all subjects where habits are to be formed at all, it is of

---

[1] See W. F. Dearborn, "Experiments in Learning," *Journal of Educational Psychology*, Vol. i, 1910, p. 373.

the greatest importance that such as are really the best should be acquired from the outset. In this way only can the necessity for unlearning be avoided.

PROBLEMS AND EXERCISES.

1. My garden gate is in the centre of a high wall. While working in the garden I hear the swish of a bicycle pass the gate. No matter how quickly I look up I never succeed in seeing the bicycle. How is this?

2. A lecturer delivering an important lecture uses his pocket-handkerchief to clean the blackboard. He remains unaware of this fact till it is pointed out to him. Explain this.

3. Describe the state of mind known as "absent mindedness."

4. A man can dodge a stone thrown at him but cannot dodge a bullet fired from the same place. Why is this?

5. A snap-shot of the start of a flat race shows the smoke of the pistol but the runners all toeing the line. Explain this?

6. If my finger actually starts the stop-watch $\frac{1}{5}$ sec. after I hear the pistol and stops the watch $\frac{1}{5}$ sec. after I see the winner breast the tape, will the timing of the race be accurate?

Are any errors introduced into the timing of a 220 yards straight race if I stand at the finish, start the stop-watch when I hear the pistol fired at the starting-line and stop the watch simultaneously with the breasting of the tape by the winner? (Sound travels 1100 ft. per sec.; and it takes $\frac{1}{5}$ sec. to start a watch after hearing a signal.)

7. How would you arrange to use a stop-watch to time 100 yds. flat race so as to avoid as many errors as possible?

8. Draw a zigzag line down a sheet of paper. Find how many turns you can make in the zigzag in 10 seconds.

9. Transliterate page 154 (*a*) into Greek letters, (*b*) into dot and dash notation.

10. Can a dog catch a biscuit dropped from only 6 inches above its nose?

CHAPTER XIV

MEMORY

The memory-image—Tests of memory—Connection between memory-
images—Associations

Apparatus and materials required.

*Tray. Cloth. A number of common objects coloured
and colourless, e.g. tumbler, fruit, chain, coloured
pencil, small book, etc. Cards with consonants clearly
marked on them for test of eye-memory. Pens, pencils,
paper, stop-watch.*

The memory-image.

Up to the present this book has dealt almost entirely
with the means by which we become aware of objects
which are at the time actually present. An attempt
has been made to show how from various sources we
gradually develop our knowledge of the world outside
us. Such knowledge could never, however, have reached
the completeness or accuracy that it has but for the
fact that we are able to bring before our minds things
which are no longer present to sense.

EXERCISE 84. *To illustrate the way in which
absent objects can be called to mind.*

Get some one to place about 20 common objects on
a tray and cover them with a cloth. Remove the cloth
for three minutes and study the tray during that time.
Then cover it again. Now try and picture the contents
of the tray. A picture recalled in this way is called

a **memory-image.** Care must be taken that it is not confused with the after-images of sight described on p. 125. One point of difference is the much more rapid fading of the after-image under ordinary conditions. Other differences will appear before the exercise is completed[1].

Write a list of all the objects you remember to have seen and add notes as to size, colour, etc. The following questions will serve as a guide to points of importance.

1.   Is the memory-image clear and sharp?

2.   Can all the objects on the tray be pictured at once?

3.   Are the pictured or "imaged" objects the same colour as the natural objects?

4.   Can the image be altered at will, e.g. can the objects be made to change positions? Or can the whole be pictured in new positions?

5.   Where does the image appear to be when allowed to come naturally?

EXERCISE 85. *To see whether memory-images can be obtained from senses other than that of sight.*

Can memory-images be obtained of sounds, scents, or touches? Try and recall some familiar sound, e.g. sounds made by a train in starting, the sound of thunder, the barking of a dog or the distant bleating of sheep.

Try also whether familiar scents can be imagined, e.g. scent of roses, smell of cheese, coffee, or of various common chemicals, such as bleaching powder, carbon disulphide, chloroform, etc.

---

[1] Cf. *Inquiries into Human Faculty*, Sir Francis Galton, p. 255, Everyman's Library edition.

Repeat in a similar way with tastes.

Can touches be recalled ? Try to recall the feeling

Fig. 49.   Sir Francis Galton

of clapping the hands, being in a hot bath or a cold
bath ; can you recall the feeling of catching a cricket
ball, kicking a football, etc.

Are impressions made by one sense more vividly or more easily recalled than those made by any other?

A test for the accuracy of memory.

An immense variety of tests have been proposed for the accuracy of memory, but perhaps the simplest is that described by W. H. Winch[1].

The method consists of showing to the class to be tested a card bearing a number of consonants arranged in rows. Vowels are avoided as otherwise nonsense syllables tend to be formed, and this affects the ease with which the letters are recalled. Any arrangement of consonants which suggests a word should be avoided for the same reason.

When it is wished to test eye-memory alone repetition of the letters is not allowed. After being exposed for a few seconds the card is covered and the person to be tested writes down all the letters which can be recalled in their proper positions. A letter in its correct position receives three marks, a letter one space away from its correct position vertically or horizontally receives two marks, a letter two spaces away one mark. No marks are given when the error is made in a diagonal direction.

To test ear-memory, i.e. the memory for heard sounds, a similar method is used but instead of looking at consonants they are called over twice in a clear voice and then the letters remembered are written down as before. Marks are given in the same way as in the last test, but of course errors can only occur in one direction.

---

[1] "Immediate Memory in School Children," *Brit. Journ. of Psych.*, 1904, (i) p. 127, and 1906, (ii) p. 52.

EXERCISE 86. *To test eye- and ear-memory by Winch's method.*

For eye-memory get your teacher to show you a card on which twelve consonants, selected as above described, are arranged in three rows of four each.   Fig. 50 shows such a card ready for use.

s	k	m	w
b	f	l	h
g	t	r	n

Fig. 50.   Specimen card for testing eye-memory

Make certain before beginning the test that the card is in such a position that it can be clearly seen.   Have it exposed for 30 secs., then remove it and write down the letters as far as possible in their correct positions. After the letters have been written down, look at the card once more and correct the results.

For ear-memory tests use similar sets of 12 consonants, but get your teacher to call them out twice over in groups of three.   Correct the results as before.

If it is intended to compare the results with those obtained by Mr Winch, the papers quoted on p. 159 should be consulted to ensure that the test is given in as far as possible exactly the original form.

Connection between memory-images.

Memory-images in the mind are unlike after-images of sight and other senses in the following respect, among others. One memory-image will, as it were, "attract" another into the mind. The following experiment will make this clear.

EXERCISE 87. *To show the connection existing between two memory-images.*

Think of some common object and then allow the mind to wander. Describe very carefully the memory-image you begin with and the one which succeeds it, and explain if you can why the two are connected.

Here are two examples which show what is required :

(i) The object thought of was a carpenter's oval pencil; after a moment or two a carpenter's bench was imagined covered with various tools, and conspicuous among them were a large plane and a shooting-board. These things were connected in the mind with the pencil because they have often been seen and used together.

(ii) The thing thought of was the word *damp*; immediately the word *lamp* flashed into the mind. In this case the objects referred to never entered the mind at all; the words were connected simply by their similarity in sound.

In these two examples, which are the results of actual experiments, we have types of two of the ways in which memories of various objects are connected with one another. In the first case the connection is between the memories of objects which have at some previous time been seen, handled or generally *sensed* near to one another in space or in time. In the second case the two experiences connected were in some way similar.

Two memory-images connected in this way are said to be associated or an **association** has been formed between them. Probably no process which takes place in the mind could be named which plays a more

important part in daily life than the power of association.
Upon it depends the use of language.   Take as a single
example the printed word DOG in English.   Directly
an educated person sees the symbols which make up the
printed word, the sound comes to mind.   This association between printed word and sound is one which is
painfully acquired within the memory of everybody and
we call the formation of such associations **learning to
read.** The formation of associations with written words
comes later and is called **learning to write.**   Much
earlier than either, in a time before most people's
memories begin, an association must have been formed
between the sound of the word and the object or collection of objects which go by the name.   At least one
other important association is formed.   When a word
spoken by some one else is heard the sensation received
is a **sound** sensation.   When on the other hand a word
is spoken by oneself a sound of quite a different
character is heard and is accompanied by sensations
of **muscular contraction** in the throat, tongue and
neighbouring parts.

In learning the use of a word not only are associations formed between object and heard word, heard
word and printed word, but also between heard word
and spoken word, printed word and spoken word, etc.

The sensations associated are roughly indicated in
the diagram below by oblongs, the directions in which
associations take place by lines.   It will be noticed that
some of the associations are used by the mind much
more frequently than others.   Thus the line joining
printed word and object is used in **silent reading,** the
line printed word to spoken word when **reading aloud**,
and so on.

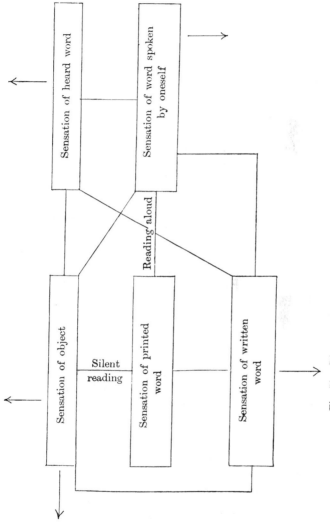

Fig. 51. Diagram showing the associations of the word DOG

11—2

Which lines represent the directions of associations which are used when

(1)   You listen to a person reading aloud.

(2)   You see a friend and *think* "There is so-and-so."

(3)   You see a friend and *say* "There is so-and-so."

(4)   You write a passage from dictation.

(5)   You copy out a piece of poetry.

(6)   You translate from French into English.

Projecting from various parts of the diagram arrows may be seen. These indicate possible lines of connection with other objects, words, etc. Thus the printed word "dog" might be associated with the printed word "dig," the sound "dog" with the sound "fog," while the object "dog" may call to mind the object "water" or "rat."

EXERCISE 88.   *To illustrate the use of association in language.*

Work out a scheme similar to the above, starting with each of the following words: "HAND," "PEBBLE," "BEAN," "TADPOLE." Try others yourself.

Trains of thought.

When the mind is left to itself and undisturbed by the entrance of any very striking sensations from the outside world, the memories of various objects, incidents, people, etc., pass before the mind in an irregular order, each of them, however, associated both with the one that precedes and the one that follows it.

No.	WORD	CONNECTION
1	Hand	
2	Arm	
3	Nail	⎫ ⎰ Thought first of nail on hand
4	Hammer	⎭ ⎱ then of nail in wall
5	Edward the First	The Hammer of the Scots
6	Edward the Seventh	Recent events have made this ready to appear
7	Coronation	
8	Scone	
9	Scotland	
10	Forth Bridge	
11	Girders	
12	Menai Bridge	Connected with no. 10, not no. 11
13	Menai Straits	
14	Anglesea	
15	Suspension Bridge	Connected with no. 12
16	Bristol	Clifton Suspension Bridge
17	Hungerford	Clifton Suspension Bridge once spanned the Thames at Hungerford
18	Germany	The name has a German sound
19	Students with spectacles	The conventional German student
20	Scoring 0 in two successive innings	

Exercise 89. *To examine the process which takes place in a train of thought.*

Allow the mind to wander freely from subject to subject and write down, or better, call out to a companion and have written down from dictation, the names of the various objects which present themselves. Continue until 20 words have been written down. Before the series has been completely forgotten read over the list and make a note of the reason why each thought suggested the next. On page 165 is a list of 20 such words representing the first ideas which came to the mind in such an experiment. In the right-hand column are notes explaining the connection in cases where this is not obvious.

Repeat the experiment several times. Also try the effect of producing a much longer series.

Also try the result of including words other than the names of things. Make a careful note of all your experiments and the results obtained.

This experiment is of little or no value unless honestly performed, i.e. unless everything thought of is written down. If it is done conscientiously a most valuable record is obtained of those ideas, or at any rate of that kind of idea, which comes frequently and easily into the mind.

PROBLEMS AND EXERCISES.

1. Write down 20 words (representing ideas) in the order in which you think of them, beginning with the word ——. Take the time with a stop-watch. When your hand is rested, find out how long it takes you to write the same 20 words from dictation. Is there any difference in the times ? Give any explanation of the result you can think of.

2. Your teacher will read out 20 words. Write down these words, and after each write down another word to represent the next idea which comes into your head.

3. Write down as many words (nouns or verbs) as you can in the time allowed (two minutes), beginning with the word supplied by your teacher. Do not begin a new word after your teacher says "stop."

4. Describe exactly what you understand by the "mind's eye."

5. In what school subjects do you find you use (*a*) sight-images, (*b*) touch-images, (*c*) sound-images, most freely for remembering facts ?

6. Do you recall a piece of music by recalling the sounds, or the feel of the keys on the piano, or their appearance on the printed score, or do you use more than one of these ? or some other form of memory ?

7. What sort of images, if any, do you use when thinking of (*a*) Numbers, (*b*) Days of the week, (*c*) Months of the year ?

8. Try and describe the train of thought which occurs in solving a simple rider in Euclid.

9. Get your teacher to read to you, slowly, an extract from some book describing scenery or events. Describe as fully as you can what you see with your mind's eye as the passage is read.

Sample passage suitable for reading :

"It was a low four-poster shelving downward in the centre like a trough and the room was crowded with impracticable tables and exploded chests of drawers full of damp linen. A graphic representation in oil of a remarkably fat ox hung over the fire-place, and the portrait of some former landlord (who might have been the ox's brother, he was so like him), stared roundly in at the foot of the bed. A variety of queer smells were partially quenched in the prevailing scent of very old lavender ; and the window had not been opened for such a long space of time that it pleaded immemorial usage and wouldn't come open now." *Martin Chuzzlewit*, ch. 31.

10. Try whether you can remember and write down seven-figure logarithms after hearing them read once.

11. A few lantern slides will be shown, each for a limited number of seconds. Draw or write down what you remember of each.

12. Take 20 words from the index of this book, beginning at ——. Hide the lower words with a slip of paper until their turns come. Write down for each the next word that comes into your head.

APPENDIX

List of Works Suitable for Reference

The following brief list of books is condensed within the smallest possible limits and consequently many very valuable works cannot be referred to.

ANATOMY AND PHYSIOLOGY.

WIEDERSHEIM AND PARKER, *Comparative Anatomy of Vertebrates.* Macmillan, 16s.

W. D. HALLIBURTON, *Handbook of Physiology.* Murray, 15s.

W. K. CLIFFORD, *Seeing and Thinking.* Macmillan, 3s. 6d.

ANTHROPOMETRICS.

G. STANLEY HALL, *Adolescence.* Sidney Appleton, London, 36s.

British Association Reports :
> *Anthropometric Investigations in the British Isles ;* Section H, Leicester, 1907, and Section H, Dublin, 1908.
> *Report of the Anthropometric Committee of the British Association,* to be obtained at the Royal Anthropological Institute, 50, Great Russell Street, London, W.C., price 1s.
> *Mental and Physical Factors involved in Education;* Section L, Sheffield, 1910, and Section L, Portsmouth, 1911, this latter partly republished in *School Hygiene,* Oct. and Nov. 1911.

SIR FRANCIS GALTON, *Inquiries into Human Faculty.* J. M. Dent's *Everyman* Series, 1s.

G. M. WHIPPLE, *Manual of Mental and Physical Tests.* Warwick and York, Baltimore, $2.50 (11s.).

JOURNALS AND PERIODICALS.

Biometrika. Cambridge University Press, price 30s. per vol.

British Journal of Psychology. Cambridge University Press, price 15s. per vol.

Journal of Educational Psychology. Warwick and York, Baltimore, 7s. 5d. per vol.

Monograph supplements dealing with Educational matters are published in connection with the above.

Journal of Experimental Pedagogy. Longmans, Green and Co., 1s. per number.

School Hygiene. Magazine and Publication Syndicate, 29, Henrietta Street, London, W.C., 6d. monthly.

PSYCHOLOGY.

C. S. MYERS, *An Introduction to Experimental Psychology.* Cambridge University Press, 1s.

E. C. SANFORD, *A Text-book of Experimental Psychology.* D. C. Heath and Co., 6s.

E. B. TITCHENER, *Experimental Psychology.* Vol. I, *Qualitative.* Part I, *Student's Manual,* 7s. net. Part II, *Instructor's Manual,* 10s. net. Vol. II, *Quantitative Experiments.* Part I, *Student's Manual,* 6s. net. Part II, *Instructor's Manual,* 10s. 6d. net. Macmillan and Co.

G. F. STOUT, *Groundwork of Psychology.* W. B. Clive and Co., 4s. 6d.

WM. BROWN, *Mental Measurements.* Cambridge University Press, 3s. 6d.

W. WUNDT, *Principles of Physiological Psychology.* Geo. Allen and Co., 12s.

C. S. MYERS, *Text-book of Experimental Psychology.* (2 vols.) Cambridge University Press, 10s. 6d. net.

STATISTICS.

A. L. BOWLEY, *Elements of Statistics.* P. S. King and Son, 10s. 6d.

W. P. ELDERTON, *Frequency Curves and Correlation.* C. and E. Layton, 7s. 6d.

W. P. and E. M. ELDERTON, *Primer of Statistics.* A. and C. Black, 2s.

C. B. DAVENPORT, *Statistical Methods.* Chapman and Hall, 4s. 6d.

INDEX